P9-DJA-793

THE COMPLETE
RF TECHNICIAN's
HANDBOOK

THE COMPLETE
RF TECHNICIAN's
HANDBOOK

By Cotter W. Sayre

PROMPT®
───────────
Publications

An Imprint of
Howard W. Sams & Company
A Bell Atlantic Company
Indianapolis, IN

FIRST EDITION, 1996

PROMPT® Publications is an imprint of Howard W. Sams & Company, a Bell Atlantic Company, 2647 Waterfront Parkway, E. Dr., Suite 300, Indianapolis, IN 46214-2041.

International Standard Book Number: 0-7906-1085-X

Library of Congress Catalog Card Number: 96-69139

Acquisitions Editor: Candace M. Drake
Editor: Natalie F. Houck
Assistant Editors: Karen Mittelstadt, Pat Brady
Illustrator: Cotter Sayre
Typesetter: Natalie Houck
Layout: Suzanne Lincoln, Phil Velikan
Cover Design: Kelli Ternet

Printed in the United States of America

9 8 7 6 5 4 3 2 1

Dedicated to my lovely wife Linda,
without whom this book would not have been possible.

CONTENTS

INTRODUCTION 1

CHAPTER 1

The Basics 3

1-1. Reactance, Impedance, and Phase Angle *3*
    1-1A. Reactance 3
    1-1B. Impedance 4
    1-1C. Phase Angle 4
1-2. Decibels *4*
    1-2A. Logarithms 7
    1-2B. Reference Levels 7
1-3. Sensitivity *7*
1-4. Semiconductors *9*
    1-4A. Diodes 9
    1-4B. Transistors 13
    1-4C. Thyristors 17
1-5. Filters *19*
    1-5A. Passive Filters 19
    1-5B. Active Filters 23
    1-5C. Crystal and Ceramic Filters 25
1-6. Toroids *26*

CHAPTER 2

Amplifiers 27

2-1. Amplifier Circuit Configurations *27*
    2-1A. Common-Base 27
    2-1B. Common-Emitter 29
    2-1C. Common-Collector 29
2-2. Classes of Operation *31*
    2-2A. Class A 31
    2-2B. Class AB 31
    2-2C. Class B 31
    2-2D. Class C 33
2-3. Amplifier Biasing *33*
    2-3A. Class A and AB Biasing 33
    2-3B. Class C Biasing 34
    2-3C. Class B Biasing 35
2-4. Amplifier Coupling *35*

2-4A. Capacitor Coupling 35
2-4B. Inductive Coupling 37
2-4C. Direct Coupling 37
2-4D. Low Frequency Transformer Coupling 37
2-4E. HF Transformer Coupling 37
2-4F. Matching Networks 39

2-5. Distortion and Noise *41*
2-5A. Distortion 41
2-5B. Noise 42

2-6. Voltage Amplifiers *42*
2-6A. RF and IF Amplifiers 43
2-6B. Preselectors 43
2-6C. Cascode Amplifiers 45
2-6D. Wideband Amplifiers 45
2-6E. Bootstrap Amplifiers 45

2-7. Power Amplifiers *45*
2-7A. Single-Ended Amplifiers 47
2-7B. Double-Ended Amplifiers 47

2-8. DC Amplifiers *51*
2-8A. Darlington Pair Amplifiers 51
2-8B. Differential Amplifiers 51
2-8C. Operational Amplifiers 53

2-9. Vacuum Tubes *54*
2-9A. Tube Types 55
2-9B. Tube Amplifier Biasing 57

2-10. Amplifier Frequency Limitations and Compensation *59*
2-10A. Frequency Limitations 59
2-10B. Frequency Compensation 59

CHAPTER 3
Oscillators and Frequency Synthesizers 61
3-1. Sine-Wave Oscillators *61*
3-1A. LC Oscillators 63
3-1B. Crystal Oscillators 65
3-1C. RC Oscillators 67
3-2. Frequency Synthesis *68*
3-2A. Phase-Locked Loop 68
3-2B. Direct Digital Synthesis 69

CHAPTER 4
Amplitude Modulation 71
4-1. AM Fundamentals *71*

4-1A. Modulation and Demodulation 71

4-1B. Overmodulation 75

4-1C. AM Power Measurement 77

4-2. AM Receivers **77**

4-2A. AM Superheterodyne Circuits 77

4-2B. Image Frequency 79

4-3. AM Transmitters **79**

4-3A. Modulation 81

4-4. AM Disadvantages **82**

CHAPTER 5

Frequency Modulation **83**

5-1. FM Fundamentals **83**

5-1A. FM Modulation 83

5-1B. Output Power and Significant Sidebands 85

5-1C. FM and AM Comparisons 87

5-2. FM Receivers **89**

5-2A. Detectors 89

5-2B. Limiters 93

5-3. FM Transmitters **93**

5-3A. Modulation Circuits 93

5-3B. Frequency Multipliers 97

5-4. Selective Calling **97**

5-5. Pre-Emphasis and De-Emphasis **99**

5-6. FM Transceiver **99**

CHAPTER 6

Single-Sideband Suppressed Carrier **103**

6-1. SSB Fundamentals **103**

6-1A. SSB Modulation 103

6-1B. SSB Output Power 105

6-1C. Standard AM and SSB Comparisons 105

6-2. SSB Receivers **105**

6-2A. SSB Detection 107

6-2B. SSB AGC 107

6-3. SSB Transmitters **109**

6-3A. Balanced Modulators 109

6-3B. Sideband Attenuation 111

CHAPTER 7

Other Communication Applications **113**

7-1. Multiplexing **113**

7-1A. Frequency Division Multiplexing (FDM) 113
7-1B. Time-Division Multiplexing (TDM) 115

7-2. Data Communications *115*
7-2A. Pulse Modulation 117
7-2B. PSK and QAM Modulation 119

7-3. Satellite Communications *121*
7-3A. Frequency Reuse 123
7-3B. The Transponder 123

7-4. Spread Spectrum *123*
7-4A. Frequency Hopping 125
7-4B. Direct-Sequence 127

7-5. Cellular Communications *127*
7-5A. The Cellular Transceiver 127

7-6. CW Communications *129*
7-6A. CW Receivers 131
7-6B. CW Transmitters 131
7-6C. CW Spurious Responses 133
7-6D. A Solid-State CW Transceiver 133

7-7. Repeaters *135*
7-7A. Radiolink 137

CHAPTER 8

Microwaves **139**

8-1. Microwave Semiconductors *139*
8-1A. Microwave Diodes 139
8-1B. Microwave Transistors 141

8-2. Microwave Tubes *143*
8-2A. Klystrons 143
8-2B. Magnetrons 145
8-2C. Traveling-Wave Tubes 145

8-3. Microstrip, Stripline, and Cavity Resonators *145*
8-3A. Microstrip and Stripline 145
8-3B. Cavity Resonators 146

CHAPTER 9

Antennas and Transmission Lines **147**

9-1. Antennas *147*
9-1A. Polarization 149
9-1B. Common Antenna Types 149

9-2. Transmission Lines *151*
9-2A. Transmission Line Types 151
9-2B. Reflected Waves 153

9-2C. Waveguides 155
9-2D. Antenna Matching Networks 155
9-3. RF Propagation *157*

CHAPTER 10
Troubleshooting **159**
10-1. General *159*
10-1A. Safety 159
10-1B. Basic Troubleshooting Steps 160
10-1C. Signal Tracing 161
10-1D. Signal Injection 163
10-1E. Lead Dress 163
10-1F. Heat Sinks 163
10-2. Intermittents *163*
10-3. Parts Failure and Their Indications *165*
10-4. Linear IC Troubleshooting *166*
10-5. Amplifier Self-Oscillation Troubleshooting . . . *166*
10-5A. Feedback Oscillations 166
10-5B. Parasitic Oscillations 167
10-6. Receiver Troubleshooting *168*
10-6A. Receiver Interference, Distortion and Noise Troubleshooting . . 169
10-6B. AM Receiver Troubleshooting 170
10-6C. FM Receiver Troubleshooting 171
10-7. Transmitter Troubleshooting *171*
10-7A. Common Transmitter Problems and Troubleshooting Techniques 171
10-7B. Transmitter Tube Final Troubleshooting 175
10-7C. Transmitter Harmonics 175
10-8. Common Tests *175*
10-8A. Transistor In-Circuit Test 176
10-8B. Transistor Out-Of-Circuit Test 177
10-8C. N-Channel FET Out-Of-Circuit Test 177
10-8D. MOSFET Out-Of-Circuit Test 179
10-8E. Capacitor Out-Of-Circuit Test 179
10-8F. SCR Out-Of-Circuit Test 179
10-8G. Diode Out-Of-Circuit Test 179
10-8H. Zener Voltage Out-Of-Circuit Test 179
10-8I. AM Modulation Percentage Test 181
10-8J. FM Deviation Test 183
10-8K. Harmonic and Overmodulation Test 183
10-8L. Amplifier Bandpass Test 183
10-8M. Phase Angle Comparison Test 185
10-8N. Resistance In-Circuit Test 185
10-8O. Receiver Sensitivity Test 185

10-8P. Receiver Frequency Response Test 186
10-8Q. SSB PEP Transmitter Test 186
10-8R. SSB Frequency Test 186
10-8S. AM/FM Output Power Test 186
10-8T. AM/FM Transmitter Frequency Test 187
10-8U. Ohmmeter Click Test 187
10-9. Test Equipment *187*
10-9A. RF Probes 187
10-9B. The Multimeter 189
10-9C. The Spectrum Analyzer 190
10-9D. The Oscilloscope 195
10-9E. The SWR/Wattmeter 199
10-9F. The Frequency Counter 201
10-9G. The Function Generator 203
10-9H. The Logic Probe 205
10-10. Soldering *205*
10-10A. Through-Hole Soldering 205
10-10B. SMD Soldering 206
10-11. Digital Troubleshooting *207*
10-12. Electromagnetic Interference *209*
10-13. Important Troubleshooting Notes *210*

CHAPTER 11

Support Circuits 213

11-1. Speech Processing *213*
11-2. Mixers *215*
11-2A. Single-Ended FET Mixers 215
11-2B. Single-Ended Bipolar Transistor Mixer 217
11-2C. Diode-Ring Mixer 217
11-2D. Autodyne Converter 217
11-2E. On-Chip (IC) Converter 217
11-3. Automatic Frequency Control *219*
11-4. Automatic Gain Control *219*
11-5. Squelch *223*
11-6. Power Supplies *225*
11-6A. Rectification 225
11-6B. Filtering 225
11-6C. Regulation 227
11-6D. Circuit Protection 231
11-7. Digital Signal Processing *231*
11-8. Miscellaneous Circuits *233*
11-8A. Comparator Amplifier 233
11-8B. Clippers, Slicers, and Clampers 233

11-8C. Voltage Multipliers 235
11-8D. Constant-Current Sources 235
11-8E. Link Coupling 236
11-8F. Electronic Switches 236

APPENDICES **237**
Appendix A: Schematic Symbols 237
Appendix B: Important Digital Gates and Circuits 238
Appendix C: Miscellaneous 239
Appendix D: Common Electronic Components 240
Appendix E: Important Reference Tables 242
Appendix F: Important Formulas 243

GLOSSARY 247

BIBLIOGRAPHY
Recommended Reading **267**

INDEX 269

INTRODUCTION

This is the handbook for the RF communications beginner, student, or experienced technician! Although meant for persons with a prior foundation in electronics, the Basics chapter will furnish the reader with a valuable review of the fundamental concepts important to the study of RF communications. This first chapter includes discussions on reactance, impedance, phase angle, decibels, reference levels, semiconductors, and filters.

All too often electronic books are loaded with theory, but contain precious little practical knowledge. As well, many of these books refer to test equipment and circuits that have not been used by industry in over twenty years. *The Complete RF Technician's Handbook*, however, includes only those circuits that you will find in the majority of modern RF devices. Additionally, current test equipment and their applications are covered in order to help the technician become familiar with their basic functions. Troubleshooting of RF, digital, and audio circuits down to the component level is heavily stressed, as this is what most competent technicians will be required to do. Since digital circuits and modulation techniques have become such an important part of many electronic devices, digital methods and their real world circuits are discussed as they apply to wireless communications. RF voltage and power amplifiers, in addition to their classes of operation, configurations, biasing, coupling techniques, frequency limitations, noise and distortion are all included, as are the major types of LC, crystal and RC oscillators. Up-to-date frequency synthesis procedures are included as well. Of course, all of the RF modulation and detection methods are explained in detail — AM, FM, PM, and SSB — as are their associated circuits, such as automatic gain control, squelch, selective calling, frequency multipliers, speech processing, mixers, power supplies, digital signal processing, and automatic frequency control circuits. Various multiplexing methods and data, satellite, spread spectrum, cellular, and microwave communication technologies are discussed. A comprehensive glossary with every need-to-know radio communications term defined is included.

The Complete RF Technician's Handbook will furnish the working technician or student with a solid grounding in the latest methods and circuits employed in today's RF communications gear, as well as the ability to test and troubleshoot transmitters, transceivers, and receivers with absolute confidence.

CHAPTER 1
The Basics

Though the reader is probably quite familiar with the basics of RF electronics, a quick review of hazy, long-ago studied principles will help in fully understanding the rest of the chapters on more advanced topics. For the electronics student or technician who is not familiar with the RF basics, this chapter is a good primer on the subject.

1-1. Reactance, Impedance and Phase Angle

Understanding the principles of reactance, impedance, and phase angle are vital for the understanding of all alternating current circuits. Reactance is the opposition to sinusoidal alternating current by inductance and capacitance. A circuit with steady DC current can not have any capacitive or inductive reactance. (Direct current also produces a magnetic field around a conductor, but this field is not changing.) Impedance is the combined opposition to both the resistive (introduced by any circuit resistance) and reactive components (physical capacitors and inductors, or stray capacitance and inductance) of a sinusoidal alternating current circuit. Phase angle is the difference in degrees between two sine voltages or two sine currents, or a combination of both; usually only of concern when working with certain amplifiers, phase splitters, angle (FM) modulation methods, and resistive and reactive components in an AC circuit.

1-1A. Reactance

There are two types of reactances; inductive and capacitive. Inductive reactance is the opposition to alternating sine-wave current that takes place in a coil or inductive circuit and increases with an increase in frequency. Inductance resists any changes in current: if the current in an inductor attempts to increase, a temporary counter-EMF, produced by the expanding magnetic lines of force that are created by the increasing current, will oppose any changes in the original voltage value. This keeps the current at its original level. The current does eventually rise, nonetheless, because the counter-EMF will no longer be induced as soon as there are no rising or falling current variations (the current has reached its maximum or minimum level); thus, no expanding or contracting magnetic fields are produced. However, if the current does begin to decrease, the magnetic lines of force cut the conductors on their way back into the conductor, creating a voltage that opposes this change in current. An inductor therefore stores current, because when the supply voltage is removed from across the inductor the magnetic lines of force contract, inducing a voltage in the inductor, creating an output current. Since an AC circuit consists of constantly changing current levels, and an inductor opposes any change in this current, this translates into a opposition to AC, or inductive reactance.

On the other hand, capacitive reactance is the opposition to alternating sine-wave current in a capacitor or a capacitive circuit. This opposition increases with a decrease in frequency. As well, the larger the capacitance of a capacitor or circuit, the lower its capacitive reactance. A capacitor resists any change in voltage and will charge to any new level of voltage as supplied by the source. If the source decreases in voltage, the capacitor will supply current to the load until it reaches this new lower source voltage. The time it takes for the capacitor to discharge or charge is dependent on its capacitance and series resistance *(t = RC)*. As an AC circuit is constantly changing voltage levels, and a capacitor opposes any variation in this changing voltage, this also translates into a resistance to AC, or capacitive reactance.

1-1B. Impedance

Impedance is the total opposition to alternating current from the capacitive and inductive reactances as well as the resistive component in a circuit. Because the reactance and resistance of a circuit are out of phase, these values cannot simply be arithmetically summed *(Z = R + X)* as in resistive circuits, but a phasor sum must be taken. First, find the reactances of each component by $X_C = 1/2\pi fC$ and $X_L = 2\pi fL$. Then, for a series circuit (*Figure 1-1*):

$$Z = \sqrt{R^2 + (X_L - X_C)^2}$$

For a simple parallel reactive circuit (*Figure 1-2*), find the total circuit current by:

$$I_T = \sqrt{I_R^2 + (I_L - I_C)^2}$$

Then, find the total impedance by $Z = V_{APPLIED}/I_{TOTAL}$. Parallel circuits with multiple complex branches (*Figure 1-3*) are even more elaborate. With real-life circuits this can begin to get very involved, as a dozen or more reactive and resistive components can make up a normal network in series and parallel combinations. Mathematics employing complex numbers (which will not be gone into here) are the normal method adopted for these calculations. Typically, however, simple direct measurements of currents and voltages can be taken, and Ohm's law would then be utilized to find the total impedance of the circuit: $Z_{TOTAL} = V_{APPLIED}/I_{TOTAL}$.

1-1C. Phase Angle

The phase difference between reactive and resistive components mentioned above refers to an inductor's voltage leading its current by 90° and a capacitor's current leading its voltage by 90°. (*Figure 1-4.*) Resistors do not have this phase or time shift between their current and voltage: When the voltage is maximum across a pure resistance, its current is also maximum. Thus, a resistance has a phase angle of 0°. A purely reactive circuit has a phase angle of ±90°, the sign depending on whether it is capacitive (-90°), or inductive (+90°). A mixed reactive and resistive circuit has a phase angle somewhere between 0 and ±90°, depending on the LCR component values.

1-2. Decibels

Decibels (dB) have a logarithmic response like that of the human ear, with 0 dB just at the threshold of human hearing. The ear responds (is more sensitive) to a change in sound intensity at lower levels than at higher levels. An increase of 4W to 5W sounds much louder than a change from 20W to 21W, yet they are both an increase of 1dB. It is therefore the power ratios that matter. (4W to 5W is an increase of 20% in power, while 20W to 21W is a increase of only 5%.) Not only is the ratio between two values used in dB measurements, but so is the $\log_{10}$ of this ratio. This was adopted to squeeze the extremes of small and large measurement values: doubling the power output of an amplifier from 50W to 100W is a 3 dB increase; quadrupling the power output from 50W to 200W is only a 6 dB increase from the original 50W. Though this quadrupling in power has occurred, it is only twice as loud to the human ear; while an increase from 50W to 400W is only a 9 dB increase from the original 50W. The power has increased eight times and is only three times as loud to the ear. As stated above, doubling the power is a 3 dB increase, while doubling the voltage or current is an increase of 6 dB. This is due to the square of the voltage conforming to the power, because when a number is squared, its logarithm is doubled, creating voltage ratios that are double the dB amounts for the equivalent power ratios. Different power ratios and the corresponding change in dB values are given in *Table 1-1*.

The formula for these decibel power calculations is:

$$dB = 10(LOG_{10}\frac{P_2}{P_1})$$

Or, when both input voltage and output voltage or input current and output current are across or through the same impedance, then the formulas

$$dB = 20(LOG_{10}\frac{V_2}{V_1})$$

and

$$dB = 20(LOG_{10}\frac{I_2}{I_1})$$

can be used. If ZIN does not equal ZOUT, then apply the formula:

$$dB = 20LOG_{10}\frac{V_2/\sqrt{Z_2}}{V_1/\sqrt{Z_1}}$$

COMPARISON FOR A CHANGE OF POWER
LEVELS TO A CHANGE IN DECIBELS

Power Ratio	dB
100	20
10	10
2	3
1.26	1
1	0
½	-3
1/10	-10
1/100	-20

Table 1-1. Comparisons for a change of power levels to a change in decibels

Figure 1-1. A series
LCR circuit.

Figure 1-2. A parallel
LCR circuit.

Figure 1-3. A complex branch
parallel LCR circuit.

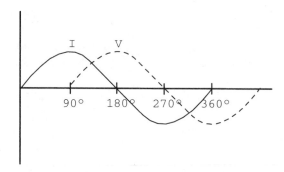

Figure 1-4. The phase shift between a
capacitor's current and voltage.

Figure 1-5. The
semiconductor diode.

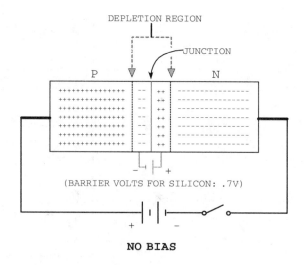

Figure 1-6. A semiconductor diode with zero
bias across its terminals.

A loss in dB, such as an insertion loss through a filter, is indicated by a negative value, while an increase in dB, such as that supplied by an amplifier, is indicated by a positive number. Either way, decibels are simply added and subtracted to find their total value.

1-2A. Logarithms

It may help to know what logarithms actually are. The log of a number is the exponent of the power of 10 that equals that number. For example, 163 equals $10^{2.212}$. Thus, the $\log_{10}$ of 163 is 2.212.

1-2B. Reference Levels

Instead of using different random values for the denominator, assorted dB reference levels have been developed. This allows comparisons to a commonly known reference level:

0dB: 6mW referenced into 500Ω: $dB = 10LOG_{10}\frac{P_{OUT}}{6mW}$

0dBm: 1mW referenced into 600Ω. Used for phone service and studio equipment in radio stations: $dBm = 10LOG_{10}\frac{P_{OUT}}{1mW}$

0dBmV: 1mV referenced across 75Ω. Used for RF signal voltage, especially on 75Ω coax transmission line: $dBmV = 20LOG_{10}\frac{V_{OUT}}{1mV}$

VU Unit: 1mW in 600Ω or 0.7746 V_{RMS} across 600Ω (or dBm units). For studio measurement in radio broadcasting.

Note: When measuring with most VOMs using the dB scale, the reference level is 0.775V for 0 dB across 600Ω.

1-3. Sensitivity

The sensitivity of an AM receiver is usually shown on the specifications table as a certain amount of microvolts (usually around 0.5 to 1μV) that will produce a signal-to-noise (S/N) ratio of 10dB. *SINAD* is another such specification, and is used for FM receivers. It refers to the ratio for *SIGNAL + NOISE + DISTORTION / NOISE + DISTORTION*, or an FM receiver's sensitivity, in μVs, with a signal-to-noise-to-distortion divided by noise-and-distortion of 12 dB at the output to the final audio amplifier.

There are two types of significant internally generated receiver noise that affect the S/N ratio, both of which produce an actual randomly fluctuating and wide frequency-ranging voltage: *white noise*, caused by a component's atoms and electrons randomly flitting around a component due to heat; and *shot noise*, caused by electrons entering the plate, collector or drain of a tube, transistor or FET, and by the random transit of electrons across a semiconductor junction. The ratio between the noise and a received signal's strength can be expressed in decibels as:

$$10LOG_{10}\frac{P_{SIGNAL}}{P_{NOISE}} \quad \text{or} \quad 20LOG_{10}\frac{V_{SIGNAL}}{V_{NOISE}}$$

and refers to a receiver's usable sensitivity.

Figure 1-7. *A semiconductor diode with the proper forward bias applied.*

Figure 1-8. *A semiconductor diode with reverse bias applied.*

Figure 1-9. *A silicon diode's characteristic curves.*

Figure 1-10. *A common small signal diode package.*

Figure 1-11. *A power diode package.*

Figure 1-12. *A popular bridge rectifier package.*

1-4. Semiconductors

In the last forty years, semiconductors have completely taken over rectification and amplification chores in RF from the once dominant vacuum tube — at least in the low-to-medium power ranges. Semiconductors are dependable, small, use low supply voltages, and are sturdy. They are used in rectification, mixing, detection, amplification, oscillation and digital circuits. ICs and most modern circuits, would be impossible without semiconductors. A quick review of their internal structure is warranted.

1-4A. Diodes

A PN junction (*Figure 1-5*) diode consists of both N and P-type semiconductor materials joined together. The N-type semiconductor has an excess of electrons (majority carriers) and few holes (minority carriers). This is due to the doping (adding of impurities) to the intrinsic (pure) semiconductor material by adding atoms that have five valence (outer shell) electrons as opposed to the four valence electrons of intrinsic silicon. P-type semiconductors have an excess of holes and a shortage of electrons in their crystal lattice structure, caused by the doping of the intrinsic semiconductor material by atoms that have three valence electrons, as opposed to the four valence electrons of intrinsic silicon. P-type semiconductor current is by hole flow through the crystal lattice. In N-type semiconductors, current is by electron flow through the crystal lattice.

With no bias voltage applied to the diode (*Figure 1-6*), electrons are attracted to the P side and holes are attracted to the N side. At the PN junction, a depletion region forms created by the combining of these electrons and holes, producing neutral electron-hole pairs at the junction, while the depletion region (on either side of the PN junction) itself consists of charged ions. This depletion region, if the semiconductor material is silicon, has a barrier potential of 0.7V. The depletion region does not increase beyond 0.7V because any more majority carriers are repelled by this barrier voltage.

If an external bias voltage of sufficient potential is applied to the PN junction of the proper polarity to forward bias the junction (*Figure 1-7*), then the barrier voltage is neutralized by this bias and electrons are able to flow. The positive terminal of the battery repels the holes and attracts the electrons, while the negative terminal of the battery repels the electrons toward the positive terminal, creating a current stream through the diode.

When a reverse bias is applied to a diode's terminals (*Figure 1-8*), the depletion region in a semiconductor diode widens as the holes are attracted to the negative side of the battery and the electrons are attracted to the positive battery terminal. With this reverse bias, the diode acts as a very high resistance, allowing very little current to flow, except for some small leakage value. This widening of the depletion region can continue until the barrier potential equals that of the bias, or breakdown occurs, creating uncontrolled reverse current flow and possible damage to the diode.

It must be remembered that approximately 0.7V will always be dropped across a forward-biased silicon diode, even when the current increases, as shown in the characteristic curves for an average silicon diode in *Figure 1-9*. This voltage drop across the diode will remain substantially at 0.7V because of the

Figure 1-13. A Zener diode.

Figure 1-14. A Zener diode's characteristic curves.

Figure 1-15. Varactor diode's schematic symbols.

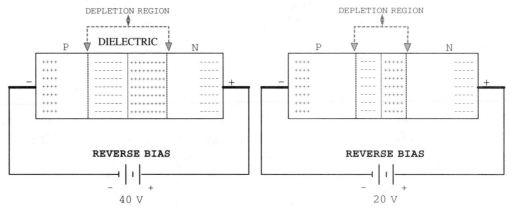

Figure 1-16. The change in "dielectric" width of a varactor diode with two different reverse bias voltages.

Figure 1-17. A varactor diode's change in capacitance versus the applied reverse voltage.

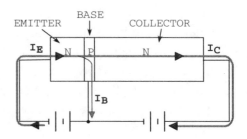

Figure 1-18. Current flow in a bipolar NPN transistor.

small value of internal (dynamic) resistance of the semiconductor materials.

Rectifier and *signal diodes* are used to perform rectification (converting AC to pulsating DC) and in mixer, RF detection and modulation circuits, as well as other circuits that need the ability of a device that conducts in a single direction only (unidirectional conduction).

Small plastic (*Figure 1-10*) and glass diode packages are used for low current applications (up to 1A). *Power diodes* (*Figure 1-11*) are capable of surviving high forward currents (up to 1,500A), while complete *bridge rectifier diode* packages (*Figure 1-12*) are used for full-wave power supplies.

Some important rectifier diode specifications:

I_R: Reverse leakage current. (The small amount of current in a reverse biased diode that increases with an increase in temperature.)

$I_{F(MAX)}$:   Maximum forward current before total destruction of a diode. Diodes can be connected in parallel to increase this current carrying capability.

PIV: Peak Inverse Voltage. (Maximum reverse voltage across a reversed biased diode.) Diodes can be connected in series to double this PIV rating.

Zener diodes (*Figure 1-13*) utilize the Zener's ability to operate with reverse bias until *avalanche*, or reverse breakdown, occurs — without being destroyed. As can be seen in the characteristic curves of *Figure 1-14*, for a large change in current, the voltage across the Zener changes very little, making it perfect for voltage regulation and voltage reference.

Zener diodes with 20, 10, 5 and 1 percent tolerances are obtainable, making it possible to use low-cost 20% Zeners in noncritical circuits, while using the higher tolerance Zeners in more precise circuits. In some circuits it must be taken into account that a standard Zener's voltage ratings fluctuate with temperature. *Voltage reference diodes*, or *temperature compensated Zener diodes*, are available that remain nearly stable with temperature variations.

Some important Zener diode specifications:

V_Z: Reverse voltage across the Zener that varies little even with an increase in reverse current.

I_{ZM}:   Maximum Zener current before Zener destruction.

P_D: Recommended maximum power dissipation rating.

I_Z: Zener current needed to keep Zener in V_Z region.

Varactor diodes (*Figure 1-15*) also use reverse bias to perform their function. They are employed as *voltage variable capacitors* (VVCs). Because increasing the width of the dielectric in a capacitor decreases its capacitance, and decreasing the thickness of the dielectric increases the capacitance, this fact is used in the varactor to electrically vary its capacitance. Increasing the reverse voltage across the varactor diode widens the depletion region, which lowers capacitance by thickening the "dielectric" (*Figure 1-16*). Lessening the reverse voltage increases capacitance by narrowing the depletion region,

Figure 1-19. A bipolar transistor's characteristic curves.

Figure 1-20. Construction of and current flow through a JFET; with schematic symbol.

Figure 1-21. The characteristic curves of a JFET.

Figure 1-22. Internal construction of an N-channel depletion-mode MOSFET with schematic symbol.

Figure 1-23. Characteristic curves of a N-channel depletion-mode MOSFET.

Figure 1-24. Schematic symbol of a dual-gate MOSFET.

thus thinning the "dielectric." *Figure 1-17* displays the capacitance variations per applied reverse voltage for a standard varactor diode. Varactors are available in numerous capacitance values for any application.

1-4B. Transistors

A bipolar (BJT) or junction transistor consists of either NPN or PNP doped regions. The *emitter* supplies the charges, while these charges are controlled by the base. The *collector* collects the charges that have not gone into the base circuit. In the silicon NPN transistor of *Figure 1-18*, we see that the emitter and the base are forward biased, while the collector is reversed biased. The emitter's electrons are repelled by the negative terminal of the battery and pushed into the thin base. The slender base cannot support the amount of electrons that accumulate there due to the small amount of holes available for recombination. Most of these electrons are therefore attracted by the positive potential on the collector, where they flow into and onto the positive bias supply, thus creating an output current. Almost all of the current supplied by the emitter flows through the collector (around 99%), while a small amount combines with the limited amount of holes in the base, then out of this base circuit. Since $I_E = I_B + I_C$ and $I_B = I_E - I_C$, then it can be seen that the currents through a transistor are directly proportional: if emitter current doubles, then so does base and collector currents. This also means that if the base current is increased by an external bias or signal, a proportional but far greater emitter and collector current will flow, creating voltage amplification if this current is fed through a high output resistance.

Since the input of a common-emitter transistor circuit is forward biased, and therefore has a low resistance, any signal injected into the base-emitter junction is across this low input resistance. This makes bipolar transistors current controlled, as shown in their characteristic curves of *Figure 1-19*. The controlling current is managed by the DC bias and external signal voltages. The signal will then add to or subtract from this DC transistor bias voltage. It must be kept in mind that transistors must also overcome their emitter-base barrier voltages, approximately 0.7V for silicon, just as in the semiconductor diode, before appreciable current can flow.

A transistor can be considered (for simplicity) as a current controlled resistance. A small current into the base controls the resistance of the transistor (which controls a much larger emitter-to-collector current), allowing this large current to flow through a high load resistance, creating an amplified output voltage.

Some important transistor ratings:

$P_{D(MAX)}$:   Maximum total power dissipation with an ambient temperature of 25°C in free air. (Can be approximated by $V_{CE} \times I_C$.)

BV_{CBO}:   Collector-to-base breakdown voltage, or the amount of collector voltage that will break down the collector junction.

$I_{C(MAX)}$:   Maximum allowable collector current.

$T_{J(MAX)}$:   Maximum internal junction temperature before breakdown of the semiconductor. Silicon has a maximum rating of around 200° centigrade.

Figure 1-25. *Internal construction of an enhancement-mode MOSFET with schematic symbol.*

Figure 1-26. *Enhancement-mode MOSFET characteristic curves.*

Figure 1-27. *Positive gate voltage creating an N-channel in an E-MOSFET's substrate.*

Figure 1-28. *Schematic symbol for the SCR.*

Figure 1-29. *How the application of varying gate current pulses affects the operation of the SCR: The higher the gate current the less VAK needed to switch the SCR on (IG1 = Low current value; IG4 = High current value).*

Figure 1-30. *Schematic symbol for a TRIAC.*

f_{ae}: β cutoff frequency, or when the frequency increases until β reaches 70.7% of its low frequency value.

f_T: Current gain-bandwidth product, or the frequency that a transistor in a CE configuration reaches a β of unity.

I_{CEO}: Leakage current between the emitter and the collector with the base open ($I_B = 0$ mA). I_{CEO} increases with an increase in the temperature and combines with the I_C, thus increasing I_C.

Junction field effect transistors (JFETs) are voltage (not current) controlled, and therefore have a very high input impedance as compared to bipolar transistors. JFETs are able to accept an input of several volts to the bipolar transistor's few tenths of a volt, while also generating less internal noise. However, FETs have less voltage gain and more signal distortion than transistors. The construction of a JFET uses a *gate, source* and *drain*. These terminals are biased so that drain-to-source voltage (V_{DS}) makes the source more negative than the drain, allowing current (ID) to flow from source to drain through the N-channel. (*Figure 1-20.*) As shown in the JFET characteristic curves of *Figure 1-21*, a JFET is a normally ON device when no bias voltage is applied. This allows maximum current flow (I_{DSS}) from source to drain. As a negative voltage is applied to the gate and source (V_{GS}), an area lacking charge carriers, called a *depletion region*, begins forming in the N-channel, and acts as an insulator. As the JFET becomes more reverse biased and increasingly depleted of charge carriers, the N-channel continues to be narrowed by the growing depletion region, which increases the channel's resistance, lowering current output into the load resistor and thus lowering the output voltage across the resistor. As the negative voltage of V_{GS} is increased on the gate, the depletion area widens, which further increases channel resistance, lowering current flow even further. A point is finally reached were the channel is completely depleted of majority carriers, and no further decrease in current is possible. This point is called $V_{GS(OFF)}$. In other words, V_{GS} effectively controls the channel resistance of the device, and thus drain current. V_{DS} (drain to source voltage) should be high enough to place the FET into a linear operating region, or above *pinch-off* (V_P). This pinch-off region is where I_D remains constant even if V_{DS} is increased — only V_{GS} can now affect I_D.

Some important JFET parameters:

Transconductance (g_m or g_{fs}, measured in Siemens or mhos): The gain of a FET, or $\Delta I_D / \Delta V_{GS}$.

V_P: Pinch-off voltage. (Minimum V_{DS} needed for linear operation of FET.)

I_{DSS}: Maximum drain current of FET possible (with $V_{GS} = 0$V).

$V_{DS(MAX)}$: Maximum allowed drain-to-source voltage.

P_D: Recommended maximum power dissipation rating.

MOSFETs (Metal-Oxide-Silicon Field-Effect Transistors) or IGFETs (Insulated Gate FETs) use a gate that is insulated from the rest of the semiconductor material, creating a device with virtually infinite input resistance. MOSFETs function in two modes: the *depletion mode*, which is a normally-on device, and the *enhancement mode*, which is a normally-off device. Depletion-mode MOSFETs (*Figure 1-22*) control drain current through the application of a negative and positive gate voltage, as shown in *Figure 1-23* for an N-channel device. Increasing the negative potential at the gate would eventually reach a point were no drain current could flow, due to the channel being depleted of all majority carriers. As V_{GS}

Figure 1-31. *Schematic symbol for a DIAC.*

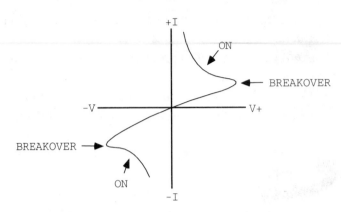

Figure 1-32. *Characteristic curve for a DIAC.*

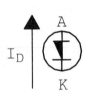

Figure 1-33. *Schematic symbol for the four-layer diode with forward current flow.*

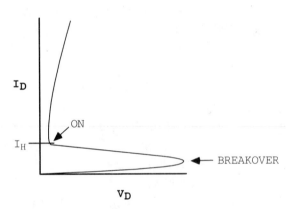

Figure 1-34. *Characteristic curve for the four-layer diode.*

Figure 1-35. *A simple LC low-pass filter.*

Figure 1-36. *A simple LC high-pass filter.*

(gate-to-source voltage) becomes less negative, more current begins to flow, until the V_{GS} reaches zero. At zero volts for V_{GS}, unlike the JFET, maximum current does not flow. The current, however, is still significant, since many majority carriers are present in the channel. As V_{GS} is made more positive in respect to the source, more electrons are drawn into the N-channel, increasing current through the channel and into the drain. Depletion MOSFETs are used extensively in RF circuits due to their low noise characteristics. Dual-gate MOSFETs (*Figure 1-24*) are used in mixers and AGC controlled amplifiers, with each gate having equal control over I_D.

An enhancement-mode MOSFET (*Figure 1-25*), as stated previously, is a normally-off device. When no bias is applied to the gate, little source-to-drain current flows, as indicated in its characteristic curves of *Figure 1-26*. A positive voltage placed on the gate creates a channel between the source and drain (*Figure 1-27*). As electrons are drawn toward the gate, an N-channel is created in the P-type substrate, allowing electrons to flow toward the positively-charged drain.

MOSFETs are easily destroyed by static electricity, as any electrical spark damages the gate insulation.

With enhancement mode MOSFETs, there is a 1V threshold voltage before any significant current will flow. They are typically used as voltage-controlled switches in applications using digital ICs.

1-4C. *Thyristors*

Thyristors are PNPN devices that are used principally as solid-state switches, and include the *SCR*, the *TRIAC*, the *DIAC* and the *four-layer diode*. They are commonly used to switch large currents, or to control the amount of power reaching a load by allowing conduction for only part of an AC alternation, or in current protection circuits.

The SCR (*Figure 1-28*) controls or switches DC (or AC half-wave) current by the application of a positive control pulse at its gate (*Figure 1-29*), which causes gate current (I_G) to flow. When the SCR latches ON by this control pulse, current then flows from the cathode to the anode and through the load. Only the lowering of the anode current (I_A) below a certain value (referred to as the *holding current* or I_H) will shut the SCR down, stopping current flow through the load. This value of holding current varies with the amount of gate current. Even with no gate pulse, the SCR can conduct when the anode-to-cathode voltage (V_{AK}) increases to a certain point, referred to as the *forward breakover voltage*. The SCR now displays very little resistance, and large amounts of current flow with only a small voltage dropped across the SCR. This forward breakover voltage point is forced to occur with less V_{AK} by increasing the gate current.

TRIACs (*Figure 1-30*) can be considered as back-to-back SCRs and perform as full wave SCRs for alternating current, conducting bidirectionally. Also, the gate control voltage, unlike the SCR, can be either positive or negative in polarity.

DIACs (*Figure 1-31*) are used as bidirectional triggers for TRIACs to equalize the ON voltage for each AC alternation. When the voltage across the DIAC's terminals reaches a certain point (the *breakover*

Figure 1-37. L-type constant-K low-pass filter.

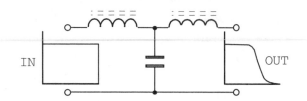

Figure 1-38. T-type constant-K low-pass filter.

Figure 1-39. Pi-type constant-K low-pass filter.

Figure 1-40. The in-series resistive component of an inductor.

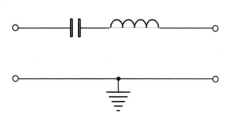

Figure 1-41. A simple series bandpass filter.

Figure 1-42. A simple bandstop filter.

voltage, *Figure 1-32*), the DIAC begins to conduct current, reliably switching on the TRIAC's gate. This breakover voltage is usually between 28V and 36V.

The four-layer diode (*Figure 1-33*, also referred to as the *Shockley diode*) switches ON and conducts current (I_D) when the anode-to-cathode voltage (V_D) reaches a certain level (the *breakover*, or switching voltage), as shown in its characteristic curves of *Figure 1-34*. The diode can only be switched off by lowering its I_D below a certain value called the holding current (I_H). A four-layer diode is capable of switching voltages up to a few hundred volts, but has a low power dissipation rating. Besides its normal current switching duties, the four-layer diode can also be used in relaxation oscillators and time-delay circuits.

Some ICs have been developed to trigger TRIACs, such as the *optocoupler TRIAC driver*, used in applications that need completely insulated TRIAC triggering with high electrical isolation.

1-5. Filters

There have been many filter designs and sub-variations over the years — entire books are written on the subject and still cannot fully cover the topic. Only the most common filters are covered here.

A filter is used to selectively pass or attenuate a certain band of frequencies, and can be *active* or *passive* using LC, RC, LCR, or LR components in the common "L", "T" or π configurations, or as completely resonant filters which can be either series or parallel (tank) resonant filters. Resonant filters enjoy widespread use in RF circuits, being used in band-pass and band-stop applications. Naturally high-Q crystals and ceramics can also be utilized to increase filter selectivity of the passband or stop-band dramatically. (The range of frequencies passed or attenuated is a function of the filter circuit's Q.)

1-5A. Passive Filters

Since the radio spectrum consists of a wide range of frequencies, some way of isolating a range of desired frequencies is needed. This can be accomplished with the use of untuned and tuned filters. A simple passive LC filter takes advantage of the ability of an inductor's increasing reactance to increasing frequencies and a capacitor's decreasing reactance as the frequency increases. An inductor is usually placed in series with the output of any generator of RF to attenuate higher frequencies, while a capacitor is placed in series with the output to attenuate lower frequencies. An inductor can also be in shunt (parallel) with the output to attenuate lower frequencies, while a capacitor can be in shunt to attenuate higher frequencies. A *low-pass filter* (*Figure 1-35*) passes only a range of lower frequencies; a *high-pass filter* (*Figure 1-36*) passes a range of higher frequencies. There are many different circuit configurations to accomplish these functions, and all may be cascaded to increase the sharpness of the skirts. Examples would be a class of filter referred to as an *L-type, constant-K low-pass filter*, as shown in *Figure 1-37*. L-type refers to its physical configuration, while constant-K refers to a filter that not only rejects certain frequencies, but matches impedances between the generator and its load throughout its entire operational passband. Other constant-K low-pass filters are the T-type (*Figure 1-38*) and π type

Figure 1-43. A tank circuit at resonance, with high internal circulating currents.

Figure 1-44. A simple parallel bandpass filter.

Figure 1-45. A simple parallel bandstop filter.

Figure 1-46. An elementary RC low-pass filter.

Figure 1-47. An elementary RC high-pass filter.

Figure 1-48. An active low-pass filter using an operational amplifier and RC components.

(*Figure 1-39*). Another common classification of filter is the *m-derived* type. This is a modified constant-K that exhibits an enhanced filter response of an extremely sharp cutoff, while displaying a constant impedance throughout its operational passband.

Other filters pass only a set range of frequencies, called a *bandpass filter*, while others reject a set range of frequencies, referred to as *bandstop, band-rejection*, or *notch filters*. These filters are almost always of the resonant or tuned type; though by combining a non-resonant low-pass with a non-resonant high-pass filter, a bandpass characteristic can be obtained.

Tuned filters work on the principle of resonance. The reactance of an inductor increases with frequency, while the reactance of a capacitor decreases with an increase in frequency. Since these are completely opposing characteristics, then when a capacitor and inductor are connected in series, a certain frequency is reached where their equal ($X_L = X_C$) but opposite reactances cancel, causing current to be maximum and impedance minimum. The only impedance to this current is the small resistance of the coil (*Figure 1-40*, with the Q of a coil being X_L / r_e), which sets the Q, and thus the bandwidth of the filter ($BW = fr/Q$). The frequency that causes $X_L = X_C$ is referred to as the *resonant frequency* (*fr*) of the tuned circuit, and depends on the values of the LC components. This *fr* can be calculated by the formula:

$$fr = \frac{1}{2\pi\sqrt{LC}}$$

Either above or below this resonant frequency, the circuit behaves as any AC circuit would to an inductor and capacitor, with the reactances now no longer canceling each other, which created a high impedance and a decrease in current flow at *fr*. As can be observed in the above formula for the resonant frequency, the larger the values of L and C, the lower the resonant frequency; while the lower the values of L and C, the higher the resonant frequency.

An example of a series bandpass is shown in *Figure 1-41*. When series resonance is reached, the current is maximum and the impedance minimum, thus passing the desired frequency without attenuation to the output. The bandstop of *Figure 1-42* has the same resonant effect, but passes the undesired frequency to ground instead of passing it to the output.

A *parallel* or resonant *tank* circuit has the opposite characteristic as that of a series resonant circuit: at *fr*, impedance is at maximum and current in the main line is minimum. This state occurs when $X_L = X_C$. As in series resonance, the value of the parallel resonant frequency can also be calculated as:

$$fr = \frac{1}{2\pi\sqrt{LC}}$$

Just above or just below this resonant frequency, the impedance decreases and the current increases. In fact, when tuning a tank, the resonant condition can be found when current is measured as minimum and voltage is measured as maximum — impedance will be extremely high. A certain amount of resistance is present in the tank circuit, lowering the overall resonant tank impedance and consuming energy, which must be replaced, creating a small current flow in the main line. Because the inductor and capacitor currents are 180° out of phase with each other, with a resonant tank having very low resistance to this

Figure 1-49. An active high-pass filter using an operational amplifier and RC components.

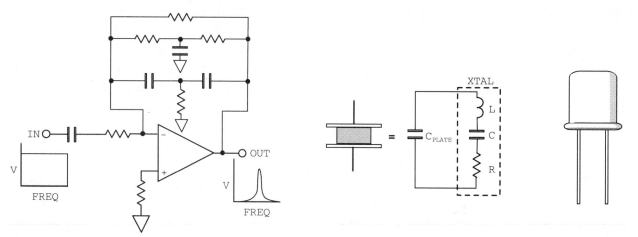

Figure 1-50. An active bandpass filter using an operational amplifier and RC components.

Figure 1-51. The equivalent circuit of a crystal in its holder.

Figure 1-52. A common crystal package.

Figure 1-53. A crystal-lattice filter arrangement.

Figure 1-54. A common ceramic ladder filter arrangement.

current, high circulating currents are present in the resonant tank, as the currents are being exchanged between the inductor and capacitor. (*Figure 1-43.*) As these currents are out of phase, however, main line current flow is minimum and dependent on the pure resistance in the tank (mainly present in the inductor). Since the main line current is at some minimal value, then impedance must be at some maximal value at *fr*. Because only at *fr* is impedance maximum, rejection of non-resonant frequencies is accomplished. Also, a complete sine wave need not be the input *fr* to achieve this resonant effect. Because L and C have the capability to store and exchange this energy and current in a circulating fashion, complete sine waves can be furnished at the output when only pulses are input. (See *Flywheel Effect.*) Only sine waves at the tank's resonant frequency are produced. The higher the Q of a resonant tank, the narrower the BW; and the higher the impedance, the higher the output voltage possible in a high Q tank. Inversely, a shunt (parallel) damping resistor can be added to a parallel tuned circuit to lower circuit Q, thus lowering gain and impedance, and increasing bandwidth.

An example of a parallel bandpass is in *Figure 1-44*. It creates a high impedance to ground for the desired frequency and produces an RF output voltage across the tank, while shunting all non-resonant frequencies. The bandstop of *Figure 1-45* passes all frequencies except the resonant frequency, which is dropped across the high tank impedance, causing a lack of output at this resonant frequency.

An RC filter is a type of passive, non-resonant filter that can be used in low and high frequency applications, or as part of the frequency determining network of an active filter. RC filters can be utilized to attenuate RF and pass only DC and low frequency AC, as shown in *Figure 1-46*. The capacitor in *Figure 1-46* has a low reactance to high frequency, but a high reactance to low frequency. Thus, the RF is sent to ground through the capacitor and the low frequencies are not. Another way to look at such a filter is to consider this circuit as a voltage divider, with the RF voltage dropped across the high resistance of the resistor instead of the low reactance of the capacitor, thus decreasing its output. Lower frequencies would cause an increase in the capacitor's reactance, dropping a larger voltage across the capacitor and less across the resistor as the frequency decreased, thus outputting increased lower frequency voltages. Swapping the resistor with the capacitor will cause the opposite effect, and create a *high-pass filter.* (*Figure 1-47.*) Any low frequencies would now be dropped across the high reactance of the capacitor and not across the lower resistance of the resistor, while the higher frequencies would easily pass through the lower reactance of the capacitor and be dropped across the high resistance of the resistor, and thus be output. Component values for both filters can be chosen to select the desired cutoff frequency.

1-5B. Active Filters

Active filters are used for audio frequency filtering. The active element is typically an operational amplifier. A sharper response curve is available over that of using passive RC components only, and eliminates the need to employ inductors, which would be too large and expensive at low frequencies. Other advantages include: lower-in-value capacitors can be adopted; the op-amp buffers the entire filter from the characteristics of the load; and the output exhibits an insertion gain instead of a loss. *Figure 1-48* shows an active low-pass filter. Due to the lower capacitive reactance of C_1 and C_2 to higher frequencies, C_2 passes the higher frequencies to ground, while C_1 sends an increasing degenerative feedback to

Figure 1-55. *A toroid used as an inductor.*

Figure 1-56. *A toroid used as a transformer.*

Figure 1-57. *A toroid employed as a transformer in the output of a broadband amplifier.*

Figure 1-58. *The popular binocular broadband transformer for semiconductor amplifiers.*

the non-inverting input as the frequency increases. C_2, R_1 and R_2 effectively form a simple passive low-pass filter.

Figure 1-49 is an active high-pass filter. R_2, C_1 and C_2 are now forming a simple high-pass filter, passing the higher frequencies to the input while the increasing capacitive reactance to the lower frequencies decreases their gain at the output.

Figure 1-50 is an active bandpass filter. The feedback network easily passes all frequencies just above and below the passband frequencies. Since this feedback is degenerative, all but this narrow band of desired frequencies is attenuated by the negative feedback.

1-5C. Crystal and Ceramic Filters

Crystal or ceramic filters are used instead of RC and LC filters because of their much higher Q and subsequent narrower bandpass and sharper response. Their frequency stability is far superior to LC circuits. Crystals perform like highly selective series and parallel resonant circuits, which have an extremely low or extremely high impedance, respectively. All crystals have a series resonant mode and slightly higher parallel resonant mode (due to the capacitance of the holder (C_{PLATE}), *Figure 1-51*). Ceramic filters are lower in Q than crystals but are also lower in cost and more sturdy — they both have almost the same characteristics. A crystal or ceramic filter uses its *piezoelectric* properties to convert mechanical vibration into electrical impulses. These vibrations only occur at the natural resonant frequency. Crystals are typically placed between a holder, inserted into a metal can (*Figure 1-52*), and marked with their series resonant frequency. The resonant frequency is dependent on the thickness of the crystal, the way it is cut, the crystal substance used, and the holder capacitance. Crystals are commonly available up to a maximum of around 25 MHz, after which their thinness makes them impractical and highly prone to stress fracture.

Crystals can be utilized alone or in combination. A *crystal-lattice filter* (*Figure 1-53*) consists of several crystals, frequently in a single package, used as a very sharp-skirted bandpass filter. Each set of crystals is cut to a different frequency to achieve the proper bandwidth and selectivity desired. The matched set of Y_1 and Y_2 have a lower resonant frequency than the matched set of Y_3 and Y_4. The difference between these two resonant frequencies sets the bandwidth of the filter. This combination of crystals passes the desired signal to the output, but blocks all other frequencies. It is used extensively in SSB communications for sideband suppression.

The *ceramic (or crystal) ladder filter* of *Figure 1-54* consists of a stack of two- or three-terminal ceramic filters. All of the crystals in this filter are cut to the same series resonant frequency, with the amount of crystals and the value of the shunt capacitors fixing the bandwidth of the filter. Coupling between individual crystals or ceramics is also accomplished by these capacitors.

Different ways of using and packaging single crystals have also been developed. A *monolithic crystal* employs quartz filters with several electrodes deposited on its top and bottom surface, with the space between the electrodes forming resonators. They can be used from 5 to 300 MHz and have a sharp frequency response: one crystal can obtain a 2 or more pole response. *Surface Acoustic Wave (SAW) filters* employ a piezoelectric crystal substrate with deposited gold electrodes. They can replace LC filters in certain wideband applications between 20 MHz to 1 GHz, and commonly experience 6dB insertion losses.

1-6. Toroids

Toroids are typically made of *ferrites*, and can function as inductors (*Figure 1-55*), transformers (*Figure 1-56*) or chokes. They have become increasingly important in RF design. Toroidal inductors and transformers have very little flux leakage, are less sensitive to inductive coupling to another coil, and can be used at high as well as low frequencies. The ferrite material can be magnetized but will not conduct electrons, thus minimizing *eddy current* loss. They do not radiate RF: unlike air-core inductors and transformers that may need shielding or changes in physical positioning to attenuate mutual coupling between components, because of their donut shape, toroids are almost self-shielding. They are also extremely efficient. For every magnetic field produced by the primary, almost all flux lines will cut the secondary. This is not true of the aircore transformer. Toroids used as interstage transformers are also supplanting aircores. They are also being placed in untuned wideband finals (*Figure 1-57*) to match the impedance between the final amplifier and the transmission line. A low-pass filter is regularly added to reduce the harmonic output, since a broadband (nonresonant) transformer or balun is incapable of attenuating these harmonics. Toroids are, furthermore, utilized to eliminate any hum that may reach the receiver from the power line, as well as attenuate any interference escaping from the transmitter into the mains.

Toroids are capable of producing small variations in inductance values by separating or pressing together their windings.

Another type of toroid transformer is the *binocular broadband transformer* of *Figure 1-58*, shown with its schematic representation. It is a common type of high- and low-power matching transformer for use in semiconductor RF amplifiers.

Broadband transmission-line transformers are also available, and can be used as baluns and as interstage matching transformers between RF transistor amplifiers. They consist, in various configurations, of a transmission line wrapped around a toroidal core.

CHAPTER 2
Amplifiers

Amplifiers can amplify voltage, current, or both. Since power is P = IE, power amplification is a natural product of this ability. Increasing either voltage or current will produce power amplification with transistors, FETs, or tubes, generating high powers into a (normally) low resistance or low impedance load.

All amplifiers employ a small external AC input signal to control a much larger DC bias current. This large fluctuating direct current is then passed into an output impedance or resistance-producing component to create a fluctuating output voltage. The output component(s) can consist of either a resistor, inductor, or tuned circuit, depending on the amplifier's intended function.

There are many ways to configure an amplifier to obtain different input and output impedances, phase inversions, frequency responses, and amplification levels. As well, negative feedback can be used for thermal stability, while various bias arrangements can produce amplification at different efficiency levels. Special coupling methods may be used to match impedances, tune-out reactances and filter out unwanted frequencies to other stages or loads.

2-1. Amplifier Circuit Configurations

There are three different circuit configurations possible with transistors (as well as tubes) and each has its own applications and abilities. They are referred to as *common-base* (grounded-grid), *common-collector* (grounded-plate), and *common-emitter* (grounded-cathode) amplifier circuits. This infers, correctly, that either the base (or grid), or collector (or plate), or emitter (or cathode) can be common to both the input and output of the amplifier circuit.

2-1A. Common-Base

The typical *common-base* amplifier configuration is shown in *Figure 2-1*. CB circuits are used as voltage amplifiers and possess low input impedance, high output impedance, and increases in output power due to $P = V^2/R$. Current gain (α or h_{fb}), however, is slightly less than unity. Any input signal is placed on the emitter, with the output taken from the collector circuit. Although the CB amplifier has great temperature stability and high frequency operational capabilities, it is rarely used except in high frequency applications due to its very low input impedance, which would normally load down any amplifier attached to its input. A positive aspect of this circuit is that it does not need neutralization because of a lack of phase shift (normally) from input to output. Common-base circuits can sometimes be found at the antenna input of a receiver because of its low Z_{IN} (50 - 75 Ω), which matches that of many antennas.

Figure 2-1. A common-base amplifier.

Figure 2-2. A popular type of common-emitter amplifier.

Figure 2-3. A common-collector amplifier.

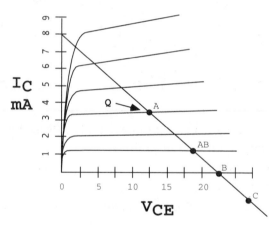

Figure 2-4. Q-point locations for the various amplifier classes.

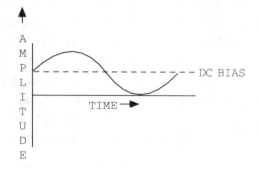

Figure 2-5. Output waveform of a class A amplifier.

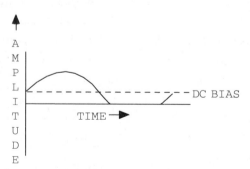

Figure 2-6. Output waveform of a class AB amplifier.

2-1B. Common-Emitter

The *common-emitter amplifier* of *Figure 2-2* is by far the most common amplifier configuration used in electronics. The type of bias shown is but one of many ways to arrange a CE amplifier (See *Amplifier Biasing*). It has the best current and voltage gain combination of any other configuration. As a signal is injected at the base, the amplified output is taken from the collector circuit. This output voltage is shifted by 180° as compared to its input signal. This shift occurs because as the base goes more positive, more current flows through the transistor, which lowers its resistance and thus the voltage across the transistor. Since the output is normally taken from the voltage drop across the transistor's collector-emitter, the load resistor drops the voltage previously available to the collector's output, creating a phase shift exactly opposite to that of the input signal. CE configurations are capable of amplifying voltage, current, and power with a medium frequency response.

2-1C. Common-Collector

One type of *common-collector amplifier* bias configuration is demonstrated in *Figure 2-3*. Also commonly referred to as an emitter-follower, the input signal is injected into the base, with the output taken from the emitter. The CC amplifier supplies current gain and some power gain. It does have a voltage gain, however, that is slightly less than unity. The main advantage of this type of amplifier is its high input impedance, which makes it valuable as a buffer or impedance matching device. The CC amplifier has a low output impedance, with a good frequency response. These amplifiers typically lack a collector resistor, and never use a bypass capacitor across the emitter load resistor, since any RF output voltage developed across R_E would then be bypassed. Also, there is no phase inversion between the input and the output signal. As the input signal to the base increases, the current through the transistor increases, increasing the current through R_E, which in turn increases the voltage drop across R_E, causing a 0° phase inversion.

Table 2-1 condenses the most important characteristics of each amplifier configuration.

COMPARISONS OF AMPLIFIER PARAMETERS

TYPE	Z_{IN}	Z_{OUT}	PHASE SHIFT
CE	Low	High	180°
CC	High	Low	0°
CB	Very Low	High	0°

Table 2-1. *Comparisons of amplifier parameters.*

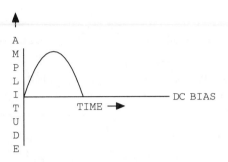

Figure 2-7. Output waveform
of a class B amplifier.

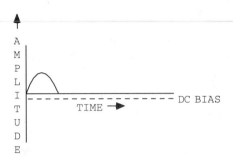

Figure 2-8. Output waveform of
a class C amplifier.

Figure 2-9. A common-emitter amplifier with
base-biased emitter-feedback biasing.

Figure 2-10. A common-emitter
amplifier showing the
Vcc connection.

Figure 2-11. A common-emitter amplifier with the
popular voltage divider, emitter-feedback biasing.

Figure 2-12. A common-emitter
amplifier with collector-feedback bias.

2-2. Classes of Operation

Different classes of amplifier operation are used to accomplish different goals. The common classes of operation are class *A*, *AB*, *B*, and *C*. Each has its own advantages and disadvantages. Every class uses circuit components to bias the active device at a different "Q", or DC operating point (*Figure 2-4*).

2-2A. Class A

Class A operation allows a signal's amplified current to flow for 360° of the input signal (*Figure 2-5*). In other words, the output never reaches saturation (the transistor conducting as hard as it can, with $V_{CE} = 0V$) or cutoff (the transistor conducting zero current, with $V_{CE} = V_{CC}$). The biasing components must bias the transistor between saturation and cutoff, or at the class A Q-point, allowing the signal to swing above and below this value. The output signal is therefore a linear and relatively accurate amplified representation of the input signal.

Class A single-ended amplifiers are generally used in low signal, non-power applications, such as in receiver RF and IF amplifiers, due to their very low efficiency. A large amount of constant DC supply power is needed at all times, since current is always flowing — with or without an input signal. They are capable of very low levels of distortion, especially with small values of negative feedback supplied by various circuit configurations.

2-2B. Class AB

Class AB operation has a slightly higher efficiency than class A, since output current flows for less than a complete cycle (around 300°, *Figure 2-6*). By lowering the Q-point, class AB operation is easily attained. This type of bias cannot typically be used as a single-ended amplifier unless a tuned circuit is used to reconstructed the entire output signal by using the *flywheel effect* (the capability of a tank circuit to complete a sine wave at the tank's resonant frequency after a pulse has been applied). Class AB is usually used for push-pull audio amplifiers, as well as in single-ended SSB linear amplifiers outfitted with the necessary output tank.

2-2C. Class B

In class B operation, output current flows for about 180° of a full cycle (*Figure 2-7*). The bias is lowered to overcome the 0.7V of the base-emitter junction only. Conduction only occurs when an AC signal alternation, or half cycle, forward biases the base. The other alternation reverse biases the emitter-base, causing a lack of output. Because the class B amplifier performs as a half-wave rectifier (amplifying only half of the incoming signal), it is typically used only in push-pull amplifier configurations (push-pull is a double-ended, or two-transistor, configuration in which each transistor amplifies each opposite alternation).

Figure 2-13. *Class A operation of a FET with source-bias.*

Figure 2-14. *Transistor operation at class C without bias.*

Figure 2-15. *Class C operation of a transistor with signal-bias.*

Figure 2-16. *A transistor with class C external-bias.*

Figure 2-17. *A transistor with class C self-bias.*

Figure 2-18. *Class B biased push-pull amplifiers.*

2-2D. Class C

Class C bias amplifies less than half of the input signal, and actually supplies only a pulse at its output port. Conduction is for 120° or less (*Figure 2-8*). The emitter-base junction is actually slightly reversed biased, or placed right at cutoff (silicon transistors, due to their 0.7V emitter-base barrier voltage, will not conduct until this voltage is overcome). Because a pulsed output is useless for most purposes, this pulse must be reconstructed into a sine wave by the use of a tuned circuit (See *Flywheel Effect*). Class C is utilized typically in non-AM RF driver stages, in single-ended non-AM RF power output stages, in AM RF modulator stages, in mixers, or as harmonic generators and frequency multipliers.

2-3. Amplifier Biasing

Since the common-emitter circuit is, as stated above, by far the most popular amplifier, we will go into more detail about its various biasing methods. Biasing is used in amplifiers for two main reasons. First, two voltage supplies would be required for the emitter-collector and the emitter-base voltages to furnish the desired class of bias (A, AB, B, or C), and to supply the necessary DC input power. Biasing was developed so that separate voltages could be obtained from a single supply. Secondly, transistors are extremely temperature sensitive, and unless a method was found to stabilize the current increase as a transistor increased in temperature, a condition called *thermal runaway* would occur, destroying the transistor. Many biasing schemes have been used to obtain these two results, with the most prevalent being the *base-biased emitter feedback,* the *voltage-divider emitter feedback*, and the *collector-feedback bias*. All three are commonly used in class A and AB operation, while class B and C amplifiers typically use other methods.

2-3A. Class A and AB Biasing

As stated above, there are three common circuit configurations employed to obtain class A or AB bias in a bipolar transistor amplifier, and their utilization depends on the desired circuit costs, stability, and other considerations. JFETs will typically obtain Class A operation with a technique called *self-bias*.

The first, referred to as *base-biased emitter-feedback* (*Figure 2-9*), functions as follows—since the base resistor, R_B, and the base-to-emitter voltage drop, V_{BE} (0.7V), and the emitter resistor, R_E, are in series with each other, and in parallel with V_{CC} (*Figure 2-10*), then as the collector current rises, due to an increase in temperature, the emitter current also increases. This increases the voltage dropped across the emitter resistor, thus reducing the voltage dropped across the base resistor (the voltage drops around a loop must equal the voltage rises). This decreased voltage across the base resistor lowers I_B, which lowers I_C. The bypass capacitor, C_E, prevents the AC signal from degenerating the gain of the circuit by bypassing the AC around the emitter resistor. R_B and R_C also allot the proper voltages to the collector and base, with the appropriate polarity, through one power supply. Considering that R_C, C-E (the collector-emitter), and R_E are in series and share V_{CC}'s voltage, then the collector-to-emitter voltage is now equal to V_{CC}, minus the voltage drop across R_C and R_E. Thus the collector is properly reversed biased. R_B, the emitter-base junction, and R_E are now also in series and share V_{CC}'s voltage. The voltage drop across R_B is equal to V_{CC} minus the obligatory emitter-base voltage drop of approximately 0.7V and the voltage drop across R_E. Considering that the voltage drops across the emitter-base and R_E are usually quite low, most of V_{CC}'s voltage is dropped across R_B, forward biasing the base. Therefore, one voltage source has now supplied the biasing needed for an NPN transistor to function as an amplifier.

Another even more effective temperature stabilization scheme than the base-biased emitter feedback method above can be employed. It is called *voltage divider, emitter-feedback biasing* (*Figure 2-11*), and works as such — the current through the voltage divider of R_1 and R_2 is much higher than the I_B. Consequently, any increase in temperature (which would increase base current) will not substantially vary V_{R2}, which is equal to the voltage at the base in respect to ground. A consistent voltage from base to ground is therefore maintained. If the emitter current increases due to an increase in junction temperature, then the top of R_E becomes more positive but, since the base is usually 0.7V more positive than the emitter, the base-emitter junction decreases in the voltage dropped across it, thus lowering collector current to its normal level. (I_C is controlled by this V_{BE} between the base and emitter.)

A biasing scheme sometimes found in circuits that need less thermal stability than above is called *collector-feedback bias* (*Figure 2-12*) and is used because it employs only two resistors, and is thus low in cost to implement. In *Figure 2-12*, as the temperature increases, the transistor will of course begin to conduct more current. The base resistor, R_B, is connected directly to the collector instead of V_{CC}. Any increase in collector current causes more voltage to be dropped across R_C, allowing less to be dropped across R_B, thus lowering the base current and, subsequently, the collector current.

A common biasing technique to obtain class A from a FET is a form of self-bias called *source-bias* (*Figure 2-13*). Unlike a transistor, no current flows in the gate circuit, so the drain current is equal to the source current. Current does flow in through the source, and thus through R_S, creating a positive voltage at the top of R_S. Inasmuch as the source is common, and the gate is at 0V in respect to ground since the gate conducts no current that would create a voltage drop across R_G, the gate can now be considered *negative* in respect to this common source. The FET is thus biased at its class A, AB, B, or C Q-point, depending on the value of the source resistor. A capacitor is usually placed across the source resistor to hold the bias voltage to a steady DC value.

2-3B. Class C Biasing

Typically, the three biasing techniques commonly used for bipolar transistor class C operation are *signal*, *external*, and *self-bias*. However, most transistors can operate class C without any bias whatsoever, such as in the circuit of *Figure 2-14*.

Signal-bias (*Figure 2-15*) uses the signal to supply the negative bias needed for class C operation. When the transistor is conducting C is charging. C then discharges through R when the signal voltage does not have the amplitude to turn on the transistor, or when the signal reverse biases the transistor. When C discharges through the resistor, a negative potential exists at the top of the resistor, thus creating the negative bias needed for class C operation. The RC time constant of the two components can allow for so much negative bias, if so desired, that only the peaks of the incoming signal can be permitted to turn on the transistor.

Another less utilized technique is external-bias (*Figure 2-16*), which employs a separate negative bias supply, consisting of either batteries or a power supply, to obtain the necessary negative bias. The RFC (Radio Frequency Choke) supplies a high impedance for the RF signal to be dropped across.

The self-bias (*Figure 2-17*) method is generally used only in applications where high input signals are available. As emitter current flows, a voltage is dropped across the emitter resistor. Since the capacitor is in parallel with the emitter resistor, it also has this same voltage across its terminals. This prevents the bias voltage from varying with the signal frequency. The top of the emitter resistor is now positive due to the direction of the current flow from emitter to collector. With the emitter now positive and being the common element in the CE amplifier, the base, which is at DC ground potential through the RFC, is now negative in respect to the emitter, creating class C operation.

2-3C. Class B Biasing

Class B, push-pull amplifiers (*Figure 2-18*), can be operated with any bias scheme that biases the amplifiers at just above cutoff. Some class B push-pull amplifiers are biased with *stabistors* (*Figure 2-19*), which are simply two series diodes employed to maintain an even 0.7V on each emitter-base junction, while also helping to protect against thermal runaway. Class B can also be used single-ended into high-Q tuned RF circuits with any biasing arrangement that biases the transistor at approximately 0.7V above cutoff.

2-4. Amplifier Coupling

A way of coupling a signal into or out of another amplifier, or a source and its load, must be used so that the DC biasing of the stages does not adversely affect each other. In many amplifiers a good impedance match for maximum power transfer (especially for power types) and the tuning-out of any reactances between stages is also critical, as well as supplying filtering and harmonic attenuation. However, most small signal cascaded amplifiers require a low output impedance and high input impedance to minimize current flow (and thus interstage loading) between stages, thus demanding mismatched impedances. The type of coupling depends on the sort of signal being amplified. DC, low frequency AC, high frequency AC, or wideband amplification all have specific coupling needs.

2-4A. Capacitor Coupling

Capacitor coupling (AKA *RC coupling*) is used in AC amplifiers only. *Figure 2-20* shows two RC coupled amplifiers. The series capacitor, C_C, blocks the DC bias to the next stage, but passes the AC signal. It can amplify large variances in frequency with a good wideband response. C_C and R_6 act as voltage dividers — since the capacitor is large enough to offer little capacitive reactance to the signal coupled to the next stage, most of the signal is dropped across the resistance of R_6 of the next stage. This adds to or subtracts from the second stage's emitter-base junction, causing its collector current to vary, producing amplification. While RC coupling is a low cost and simple method, it is, however, very difficult to match impedances from one stage to the next. It also has no attenuation effect on higher undesired frequencies, such as harmonics, from passing from one stage to the next and being amplified.

Figure 2-19. Class B push-pull amplifiers biased with stabistors.

Figure 2-20. Two CE amplifiers utilizing capacitor coupling between stages.

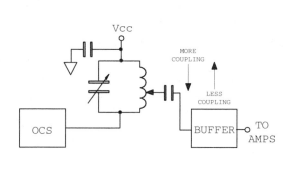

Figure 2-21. Adjustable capacitive coupling between two stages.

Figure 2-22. Two CE amplifiers utilizing inductive coupling between stages.

Figure 2-23. Two CE amplifiers utilizing direct coupling between stages.

Figure 2-24. Two CE amplifiers utilizing low frequency transformer coupling between stages.

Another type of capacitive coupling can be used in some circuits and operates in conjunction with an inductor. The degree of coupling between an oscillator and its buffer (*Figure 2-21*) can be controlled by a movable tap. Coupling is increased if the tap is moved toward the bottom, decreased if moved toward the top. Taps are utilized to decrease loading on an oscillator stage, or to increase or decrease the output signal to the buffer.

2-4B. Inductive Coupling

Inductive coupling (AKA *impedance coupling, Figure 2-22*) is also utilized for alternating current only. It is similar to RC coupling, but instead of employing a collector resistor, a *collector inductor* is used. Since the collector inductor has very little DC resistance, insignificant DC power is consumed in the inductance, allowing for more efficient operation of the amplifier. It is only usable for a narrow band of frequencies, however, since X_L varies directly with frequency. The stage gain would thus fluctuates with any change in X_L. (This is not to be confused with *transformer coupling*.)

2-4C. Direct Coupling

Direct coupling (AKA *DC coupling*) is useful for low frequency and DC amplification. *Figure 2-23* shows a typical DC amplifier. R_3 is used as a collector resistor for Q_1 and as a base resistor for Q_2. A small temperature-induced current change in Q_1 is amplified by Q_2, so DC amplifiers must be well designed, using precision components, to stabilize the circuit against these temperature variations. DC amplifiers are perfect for ICs, since all the components are placed close together, allowing each component the same changes in temperature.

2-4D. Low Frequency Transformer Coupling

Low frequency (usually laminated iron-core) *transformer coupling* is used for audio AC but not DC. The DC, of course, cannot pass through the transformer. *Figure 2-24* shows two amplifiers coupled by an iron-core transformer. The transformer can be used to match the high output impedance of Q_1 to the low input impedance of Q_2. One end of the secondary is attached to the base, with the other end attached between its bias resistors R_5 and R_6. The signal then adds to or subtracts from the base bias, causing the base current to fluctuate, creating amplification at the output of Q_2. A transformer operated in this way is called an *interstage transformer*. An iron-core audio transformer is heavy and expensive, so this coupling method is now rarely used for normal audio applications.

2-4E. HF Transformer Coupling

High frequency (tuned circuit, ferrite, powdered-iron, and air core) *transformer coupling* (*Figure 2-25*) is still used extensively in RF and IF amplifiers, but less so than in the past because of its high expense as compared to other coupling methods. Its operation is the same as iron core transformer coupling, but transformers can be air core for low losses at high frequency, or ferrite and powdered iron cores for inductive tunability and high inductance operation over and above that of the air core. The tool used to adjust the reactance of the ferrite or powdered-iron slug types is shown in *Figure 2-26*. Transformers

Figure 2-25. Two CE amplifiers utilizing high frequency transformer coupling between stages.

Figure 2-26. Plastic inductor and transformer tuning tool.

Figure 2-27. The common toroidal transformer.

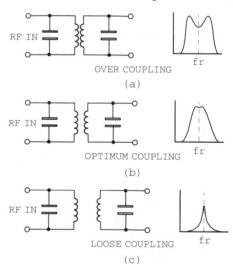

OVER COUPLING
(a)

OPTIMUM COUPLING
(b)

LOOSE COUPLING
(c)

Figure 2-28. How coupling degree between RF transformer windings affects the signal.

Figure 2-29. A high-to-low impedance matching L-network.

Figure 2-30. A low-to-high impedance matching L-network.

also supply the needed impedance matching for efficient maximum power transfer between stages. Either the primary or secondary, or both, can be tuned, forming a resonant tank circuit, to allow the coupling to be frequency selective. *Figure 2-25* shows *double-tuned transformer coupling*, allowing the tank circuit to pass just the desired frequencies while attenuating all others outside of its bandwidth. If only the primary of T_1 were tuned, then it would be referred to as *single-tuned transformer coupling*. *Toroidal transformers (Figure 2-27)* are very popular in RF gear and have, in many applications, replaced the normal slug and air core transformers.

With RF and IF interstage transformers, the degree of coupling between the primary and secondary can be a large consideration. This is usually governed by the distance between the primary and secondary windings. As the coefficient of coupling increases (*overcoupling, Figure 2-28a*), more flux lines from the primary are able to cut the secondary, creating a higher output voltage over that of *loose coupling* (*Figure 2-28c*), as well as an increasing bandwidth as adjacent frequencies are also passed. As the coefficient of coupling is decreased toward loose coupling, the amplitude as well as the bandwidth of a signal will decrease. However, loose coupling is sometimes utilized because harmonics are decreased owing to its lower capacitive coupling into the next stage, as well as supplying the narrower bandwidth needed for some applications. At *optimum coupling (Figure 2-28b)*, the bandwidth is wide enough to pass a regular narrow bandwidth signal, with an amplitude that is equal to overcoupling.

2-4F. *Matching Networks*

Various *RF matching networks* can be used between stages or their load to perform coupling, without the use of a transformer, and supply the necessary impedance matching, nullify any reactances (making the input impedances to and from a stage into a pure resistance), and furnish a certain amount of filtering. Maximum power transfer and attenuation of harmonics between stages or their load is consequently accomplished. Most modern RF equipment is now designed with matching networks rather than using the more expensive transformer coupling methods. This impedance matching can be attained by simple *L-networks*, which also supply low-pass filtering to reduce harmonic output. The L-network of *Figure 2-29* matches a high output impedance amplifier to a low input impedance amplifier. Its components values are set so that at the amplifier's operating frequency the circuit appears to be a parallel resonant circuit to amplifier 1. Since a parallel resonant circuit, at resonance, has an extremely high impedance, the two stages are matched. The L-network of *Figure 2-30* shows a low output impedance amplifier matched to a high input impedance amplifier. The inductor and the capacitor at the amplifier's design frequency appears to amplifier 1 as a series resonant circuit. A series resonant circuit, at resonance, has a very low impedance, so the two circuits are matched.

One of the most popular impedance matching networks between a low output impedance amplifier to a higher input impedance is the low cost and small *T-network (Figure 2-31)*. The T-network can be designed to supply any desired selectivity and impedance matching, within limits, that is not possible with the L-network. The capacitors are usually tunable for maximum output power during alignment.

A common tube and transistor circuit impedance matching network is the π *configuration*. The π and its equivalent circuit is shown in *Figure 2-32*. By changing the ratio between C_1 and C_2, Z_{OUT} of a power RF

Figure 2-31. *An impedance matching T-network.*

Figure 2-32. *The popular pi-matching network and its equivalent circuit.*

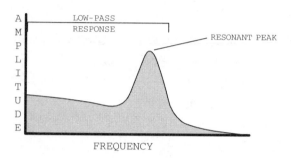

Figure 2-33. *The frequency response of a pi-network.*

Figure 2-34. *Frequency-domain graph showing IMDs.*

Figure 2-35. *Frequency-domain graph showing harmonics.*

Figure 2-36. *An ordinary RF amplifier: (a) as shown in some schematics; (b) redrawn.*

amp can be matched to the Z_{IN} of an antenna, as well as attenuating any harmonics and performing the needed flywheel function. Although the shunt capacitor π-network is also considered to be a low-pass filter, it has a resonant peak at a certain design frequency where the impedance of the circuit rises drastically above that of its low-pass value (*Figure 2-33*). While C_1 is referred to as the *tuning capacitor* and tunes the plate or collector to resonance, and C_2 is called the *loading capacitor* and varies the impedance for proper power transfer into the antenna, both heavily interact with each other. (If C_1 is adjusted, then C_2 must be readjusted until a desired compromise is reached between tuning and loading.)

2-5. Distortion and Noise

Distortion and noise are an integral part of any amplifier, mixer, or oscillator used in a receiver or transmitter. Distortion deforms either the carrier and any sidebands at the transmitter, causing adjacent channel interference and inaccurate reproduction of the original modulating signal at the receiver or, if it occurs within the receiver, creates undesirable, distorted output. Distortion can be created by the production of harmonics of the complex modulated and carrier waves, or by internal mixing in any nonlinear components (creating unrelated distortion products of varying frequencies), or by improper amplifier responses to the desired or undesired frequencies. Distortion may have different causes, but similar results — a waveform that is changed in shape or amplitude from that intended by the designer.

Noise also can result in anything from annoying white static noise to total obliteration of the received signal, originated by many internal and external causes.

2-5A. Distortion

Frequency distortion, measured as frequency response, is distortion which can occur if a circuit boosts or attenuates certain frequencies better than others, commonly a problem in audio amplifiers when they are near their extreme lowest and highest frequency limits. It is caused by the reactive components in a circuit acting in an undesirable fashion as bandpass, bandstop, high-pass, or low-pass filters, or by the inability of a transistor to perform at high frequencies, induced by transit time problems and the negative effect of junction capacitance.

Amplitude distortion, often referred to as *nonlinear distortion* because the output signal is not a true replica of the input signal, is created by improperly biasing an amplifier, resulting in the input signal causing either saturation or cutoff. This produces harmonics and IMDs (intermodulation distortion). The harmonics and IMDs generated from the resulting *flattopping* of the signal create a raucous, coarse output that can be heard in an audio amplifier. Sometimes referred to as *overload distortion*, the input signal overdrives the amplifier, usually creating both saturation and cutoff conditions. In a transmitter, amplitude distortion will cause interference to other channels.

Intermodulation distortion (IMD, *Figure 2-34*) is caused when frequencies, not harmonically related to the fundamental, are created by the mixing together of the carrier, any harmonics, the sidebands, the modulating audio, IMDs from other stages, etc., through nonlinearities in a linear or nonlinear amplifier (such as a class C) or in the output of any mixers. This produces an abundance of spurious responses both in and out of band. It can also originate in the finals of a transmitter when another nearby transmitter's signal and/or its harmonics reaches the power amplifier. One signal will modulate the other, resulting in nonlinear mixing of these signals in the power amplifier, creating many distinct sum-and-difference frequencies that are not harmonically related to the fundamental. This type of IMD can be attenuated by using a wavetrap tuned to the second transmitter's frequency. IMDs of this variety can also occur in any stage of a transmitter if the transmitter is not well shielded and grounded. Receiver IMDs, consisting of the desired signal and a nearby transmitter's undesired signal (and/or its harmonics) can be admitted into the front end of a receiver. This can cause reception of unwanted signals and a smearing and/or obliteration of the desired frequency. A notch-filter (if the interfering signal is far enough away from the desired signal) can be effective, as can confirming that all amplifiers are biased as designed, and are not functioning in a nonlinear region due to a component defect or stage misalignment. Only IMDs near the desired frequency cause problems — any created outside of the passband of the transmitter or receiver are extensively attenuated by an amplifier's tuned circuits and the selectivity of the antenna.

Harmonic distortion, measured as *total harmonic distortion (THD)*, takes place when undesired harmonically related frequencies are produced in the amplifier and add to and distort the fundamental. Interference to receivers tuned to many megahertz away from a transmitter's carrier can be created when these harmonics are broadcast into space (*Figure 2-35*). Since harmonics are created by amplitude distortion in an amplifier, the stronger the desired input signal the higher in amplitude are the undesired harmonics generated. An extreme case of distortion can result in the sine wave carrier turning into a rough square wave (square waves contain the fundamental frequency plus many odd harmonics). Because no amplifier can be completely linear, a certain amount of harmonics are created in all amplifiers and they must be suppressed as much as possible — especially in a transmitter.

2-5B. Noise

There are two major classifications of noise: *circuit generated* and *externally generated*. Both are unavoidable and limit the possible gain of any receiver's amplifiers.

Circuit generated noise causes an actual randomly fluctuating and wide frequency ranging voltage, created by two main causes: *white noise*, produced by a component's electrons aimlessly flitting around due to heat; and *shot noise*, caused by electrons entering the plate, collector, or drain of a tube, transistor, or FET, and by the random transit of electrons across a semiconductor junction.

*External noise*s are those such as static caused by atmospheric disturbances like lightning, as well as *space noise* created by solar flares and sunspots, and *cosmic noise* induced by the stars radiating wide-ranging interfering signals. *Man-made noise* also contributes to external noise generation, and can be from any type of electromagnetic interference; from dimmer switches, to car ignitions, to power tools.

2-6. Voltage Amplifiers

Voltage amplifiers are used to increase low signal amplitudes to usable levels. Receiver RF and IF amplifiers are voltage amplifiers, typically class A for linear operation and low distortion in AM and

SSB. An RF stage must ordinarily be tunable and capable of supplying relatively high gain at high frequencies, while being low in noise generation (since any noise present will be amplified by later stages), and of sufficient bandwidth to pass the desired signals and attenuate all others. Because of the high operating frequencies, RF amplifiers are frequently *neutralized* to counterbalance any possible positive feedback and its resultant self-oscillations. An IF amplifier must be high gain, since the sensitivity of a receiver is largely governed by the gain of the IF stages. They are fixed tuned so as to be highly selective and pass only the desired signal. High gain and narrow selectivity are also easier to accomplish by making the IF lower in frequency than the incoming RF.

Voltage gain of these amplifiers can be calculated as V_{OUT}/V_{IN}. With a series of cascaded amplifiers, their gain is multiplied. For example, if each cascaded amplifier stage supplies a voltage gain of 100, and there are three stages of amplification with the same gain, then the combined value would be 100 x 100 x 100 = 100^3, or one million.

2-6A. RF and IF Amplifiers

One type of widespread RF or IF stage is shown in *Figure 2-36a*. It may be easier to identify if redrawn as in *Figure 2-36b* as the common-emitter, voltage divider, emitter feedback class A voltage amplifier with tuned circuit RF transformer coupling. Collector bias, base-bias with emitter feedback, and inductive coupling or LC matching networks can be employed instead of the above configuration. An AGC connection is normally applied at the base to supply gain control. (See *Automatic Gain Control.*) The input can also be tuned in an RF amplifier, and is almost always tuned at the input of an IF amplifier. Class A biased low noise FETs are normally used in RF amplifiers at VHF and above. When increased selectivity is preferred, a crystal filter can be placed in front of the IF strip, just after the mixer, to supply the desired narrow bandwidth signal.

IC IF amplifiers are also quite common (*Figure 2-37*). With only a few external components to supply decoupling, coupling, and resonant tank functions, an entire IF stage or section can be placed upon a single chip.

Dual-gate MOSFET amplifiers, such as *Figure 2-38*, are employed as IF stages to supply very high input impedances (to reduce loading of the prior circuit), as well as the inclusion of a *control gate* to manage the gain of the stage through AGC action.

2-6B. Preselectors

A *preselector* is a tuned radio frequency preamp (or tunable passive filter, *Figure 2-39*) placed before the RF amplifier of a receiver to increase selectivity, sensitivity, S/N (signal to noise) ratio, and to attenuate any image frequency. When changing the received frequency, some preselectors must be adjusted manually by a separate control, while others are ganged with the tuning capacitor.

A simple, untuned radio frequency amplifier in a receiver is sometimes referred to as a preselector.

Figure 2-37. An IF amplifier in integrated circuit form.

Figure 2-38. A dual-gate MOSFET amplifier with one gate used for the input signal and the other for AGC.

Figure 2-39. A passive RF preselector.

Figure 2-40. The wideband RF cascode amplifier.

Figure 2-41. A shunt damping resistor (RP) used to decrease tank Q.

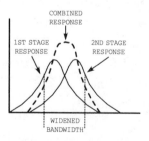

Figure 2-42. The effects of two-stage stagger tuning on bandwidth.

2-6C. Cascode Amplifiers

Cascode amplifiers (*Figure 2-40*) are low noise, high gain, high input impedance, wide bandwidth amplifiers sometimes used as a preamplifier for receivers. It consists of two transistors or MOSFETs, with the input stage acting as a common-emitter circuit feeding the second transistor, which is a common-base. If the wide bandwidth supplied by the cascode amplifier is not sufficient, other methods can be used.

2-6D. Wideband Amplifiers

Since $Q = R_P / X_L$ and $BW = f_r / Q$, a *shunt damping resistor* can be inserted into a tank circuit to lower its "Q" to effectively widen an amplifier's BW. (*Figure 2-41.*) The lower the value of R_P, the wider the bandwidth but the lower the output voltage. In most RF communications circuits, however, a narrow bandwidth is desired to attenuate adjacent channel interference and improve the S/N ratio. TV and certain wideband communication applications use this method, as well as a technique called *stagger tuning*, to further widen the bandwidth. Stagger tuning uses two or more stages tuned to a slightly different frequency (*Figure 2-42*), and offers steep skirts and better gain than merely adding a parallel damping resistor (R_P).

2-6E. Bootstrap Amplifiers

The *bootstrap circuit* (*Figure 2-43*) is an amplifier in a common-collector configuration that increases its own input impedance but, as with all CC amplifiers, has a voltage gain of less than one. A common-emitter bootstrap circuit can also be used for its increased input impedance, but voltage gain is much lower than a normal CE amplifier, and the bias (temperature) stability is compromised. The bootstrap circuit can be used with any type of active device.

2-7. Power Amplifiers

Power amplifiers (PAs) are designed to supply high power outputs into a low impedance load. The loads are low impedance because speakers are usually 8 Ω, antennas 50 to 300 Ω. Power amplifiers must be able to handle large signal inputs without saturating or cutting-off (*Figure 2-44*), while controlling large amounts of output current (but generally at low values of stage gain).

There are two fundamental classes of power amplifiers to increase RF output power for transmitters, employing either a single active device (single-ended) or two active devices (double-ended or push-pull). These two classes can be broken down into *linear* (class A, class AB and B push-pull) and *nonlinear* (class C, almost always into a tank circuit) amplifiers. Only linear RF amplification must be used for SSB and low-level AM to maintain their modulation envelopes. Class C is too nonlinear to sustain these envelopes, and severe distortion will result. As well, audio power amplifiers must always be operated in one of the linear modes, or severe distortion will occur to the audio.

Figure 2-43. *The high input impedance bootstrap amplifier.*

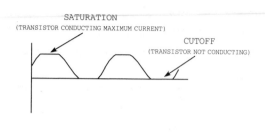

Figure 2-44. *An overdriven amplifier's waveform showing saturation and cutoff.*

Figure 2-45. *An audio power amplifier in integrated circuit form.*

Figure 2-46. *A single-ended final for low power applications.*

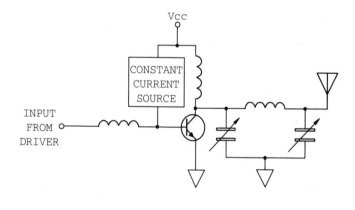

Figure 2-47. *A special single-ended final capable of supplying medium output power.*

Figure 2-48. *A class C RF driver amplifier.*

Linear IC audio power amplifiers (*Figure 2-45*) are also available for low power receiver applications where low input power consumption and voltage are critical, such as in battery operation. In the IC power amplifier example of *Figure 2-45*, using, or not using, an external resistor and capacitor between pins 1 and 8 can easily and simply vary its output gain to any desired value between 20 and 200.

2-7A. *Single-Ended Amplifiers*

The class A single-ended RF power amplifier of *Figure 2-46* can be employed in very low powered transmitter finals. A single-ended power amplifier can be quickly identified by the very low value bias resistors used. A special class A single-ended amplifier biased by a temperature compensated constant-current source can deliver up to 100W. (*Figure 2-47*.)

Class C single-ended amplifiers can be used to amplify any signal that does not contain a modulation envelope, which would be distorted by its insufficient linearity. *Figure 2-48* is an ordinary class C RF power driver stage. The transistor is operating class C without any bias, as most unbiased transistors act as class C amplifiers. A low value R_E keeps the transistor stable during temperature variations by supplying degenerative feedback. If R_E were not bypassed by C_{RE} further negative feedback would occur, lowering gain. A ferrite bead or beads can be added to the base lead to suppress any VHF and UHF parasitics as well as high frequency harmonics. RFC and C_1 are a decoupling network for the power supply, while C_2 is a tunable capacitor that resonates with the primary of T_1 to form the tank circuit. Class C amplifiers can also be used in single-ended RF power output stages, as FSK, FM, and CW amplifiers, or as harmonic generators and frequency multipliers. They are known for their very high efficiency of operation.

2-7B. *Double-Ended Amplifiers*

A common power amplifier is the double-ended, or *push-pull* type — especially for audio applications. Push-pull amplifiers use two transistors, with each transistor amplifying each opposite half of the incoming signal. The two amplified halves are then assembled at the output. There are two basic ways to obtain push-pull operation. The first uses a technique to take the input signal and split it into two 180° out-of-phase signals (*Figure 2-49*) and feed this output into the bases of two identical active devices. There are several ways to accomplish this phase splitting. The first is by using a passive center-tapped transformer phase-splitter (*Figure 2-50*). The center-tapped transformer functions as a phase-splitter because the signal between the center tap and the *top* of the transformer is 180° opposite in phase (polarity) to this center tap and the *bottom* of the transformer. The two out-of-phase signals are needed so that as one transistor is increasing its current flow, the other stage is decreasing its current flow.

Class B push-pull is a common way for power amplifiers to be configured, due to their very high efficiency over that of class A. Class AB is also frequently used in push-pull power amplifiers that require less distortion, but with less efficiency than class B. Class C RF push-pull is used for signals that do not need perfect linear amplification — a tuned circuit at the output reconstructs the input signal by the *flywheel effect*.

Figure 2-49. *A push-pull amplifier with phase-splitter.*

Figure 2-50. *A center-tapped transformer used as a phase-splitter.*

Figure 2-51. *A push-pull amplifier with active input phase-splitter.*

Figure 2-52. *A paraphase active phase-splitter circuit.*

Figure 2-53. *A transformerless push-pull amplifier.*

Figure 2-54. *A broadband RF push-pull amplifier.*

Regular push-pull amplifiers also cancel hum from the power supply and self generated even-order harmonics (these currents are out of phase and cancel in the output transformer), and can supply large currents to the load, thus producing large output powers.

Push-pull, transformerless input power amplifiers have been developed. With the use of an active phase-splitter, the circuit of *Figure 2-51* needs only a center-tapped *output* transformer, and can dispense with the expensive and bulky center-tapped *input* transformer. The *paraphase* amplifier (*Figure 2-52*) is a common and simple way to accomplish this active phase splitting. By using identical values for the collector and emitter resistors, an equal amplitude but 180° phase shifted output is produced. The paraphase amplifier can thus be used to supply the necessary out-of-phase signals to the input of a push-pull amplifier. The output transformer is operated to reconstruct the alternate outputs from Q_2 and Q_3. When Q_2 conducts, current will flow through the top of the coil; when Q_3 conducts, a opposite current will flow through the bottom of the coil, inducing an opposite voltage into the transformer's secondary, creating a complete low-distortion signal for the speaker coil. Keep in mind that these opposite currents occur at different periods in time, due to the phase shift created at the input, so they do not oppose each other, but are merely "assembled" at the output transformer. The output transformer also functions as an impedance match for the output of the audio amplifier to the low impedance of the speaker coil, as well as isolating the speaker from the DC current flow.

Completely transformerless audio push-pull amplifiers, referred to as *complementary symmetry* with *stacked-series output* (*Figure 2-53*), are designed to eliminate the expensive and heavy (at audio frequencies) input and output transformers. By using evenly matched NPN and PNP transistors, no phase inversion is necessary at the input, since they both naturally conduct at opposite times. The output is taken from the emitters in a common-collector configuration. No voltage gain is possible, but this setup does supply a high driving current into a low impedance load. D_1 and D_2 (referred to as *stabistors* when used in this configuration) keep an even 0.7V on each emitter-base junction, which consequently biases the transistors just above cutoff (class B), while also helping to protect against thermal runaway. R_3 and R_4 aid in thermally stabilizing the two transistors.

Push-pull wideband class B and AB RF amplifiers, that need not be tuned by the operator, are commonly available that amplify signals between 2 and 30 MHz (*Figure 2-54*), and are similar in operation to the audio push-pulls. These are simply broadband push-pull amplifiers without tuned input or output circuits. They function due to the phase shifting abilities of a center-tapped transformer. The phase between the top of the tapped secondary of T_1 and the bottom of T_1 are 180° out of phase — the upper stage amplifies the positive alternation of the incoming signal, while the lower stage amplifies the negative alternation. At the primary of T_2 the two signals are added and the signal is outputted through the secondary to the antenna. T_1 is also supplying an impedance match between the output of the driver stage and the input to the push-pull power amplifier, while T_2 is supplying the output of the power amplifier with an impedance match for the input to the antenna. D_1 and R_1 biases the two transistors right at cutoff, for true class B operation, to eliminate any crossover distortion that would result if the input signal had to overcome the 0.7V of the base-emitter junction before they could conduct.

An *RF parallel power amplifier* (*Figure 2-55*) is used to double the radio frequency output power over that of a single-ended power amplifier. In this configuration, up to several hundred watts output power

Figure 2-55. A parallel power amplifier circuit.

Figure 2-56. A parallel final power amplifier showing bias components.

Figure 2-57. A basic single stage DC amplifier.

Figure 2-58. Darlington pair DC amplifier.

Figure 2-59. The differential pair DC amplifier arrangement.

Figure 2-60. The differential amplifier in a single-ended input configuration.

can be obtained with power transistors. Parallel amplifiers differ from push-pull in that each transistor is on or off during the *same* time interval, unlike push-pull, which alternately shares the power work load back and forth for equal time periods. The parallel configuration, since the output current load is shared equally amongst the transistors or tubes, as it is in any parallel circuit, doubles the power handling capabilities (P = IE) over that of a single-ended amplifier. This arrangement allows the circuit to act as if it were a single high power transistor with two, or more, perfectly matched transistors increasing the output power handling ability of a transmitter's final. *Figure 2-56* shows a complete biased circuit, with impedance matching and filtering being accomplished by the π-network at its output. Parallel amplifiers can be biased class A, AB, B or C.

2-8. DC Amplifiers

Direct current amplifiers are used to amplify DC and low frequency AC currents or voltages. A simple single stage amplifier (*Figure 2-57*) without coupling capacitors can be used if high gain is not needed. If more than one stage of amplification is required then AC coupling methods, which utilize transformers, inductors, or capacitors, cannot be used in a DC amplifier. The DC signal would be blocked or shunted by these components. Other techniques must be employed.

2-8A. *Darlington Pair Amplifier*

One type of DC amplifier is the *Darlington pair amplifier* of *Figure 2-58*, configured as a high voltage-gain common-emitter circuit. Darlington common-collector amplifiers are also popular. The pair work as one transistor with an h_{fe} x 2, or $\beta = \beta_{Q1}$ x β_{Q2}. In practice, the current is not quite doubled due to other constraints, such as bias voltage adjustments and impedance mismatching. R_1 and R_2 are used for voltage divider bias, R_3 is the collector resistor, with Q_1 controlling the conduction of Q_2. When a DC or AC signal is placed at the input, the base current of Q_1 (I_{BQ1}) increases, if DC, or varies, if AC. The emitter current of Q_1 (I_{EQ1}) then increases or varies, which increases or varies the base current of Q_2 (I_{BQ2}), which increases or varies the collector current of Q_2 (I_{CQ2}), creating amplification. Notice that no R_E is used due to the degenerative action that would result in substantially lower gain. Darlingtons can be placed in the same three terminal package as a standard bipolar transistor.

2-8B. *Differential Amplifiers*

Another popular DC amplifier is the *differential pair amplifier* (*Figure 2-59*), which provides great thermal stability and high gain. It is now typically manufactured in integrated circuit form and rarely as a discrete circuit. This circuit is usually configured to amplify the voltage *difference* between the V_1 and V_2 outputs.

The differential amplifier is versatile in that there are two basic ways that it can be configured. For *differential output*, both inputs are fed a signal with the output taken across V_1 and V_2. (See *Figure 2-59*.) Amplification will only take place if the input signals are either out of phase or of different amplitudes. With equal or *common-mode signals* at the inputs, they will not be amplified since a difference

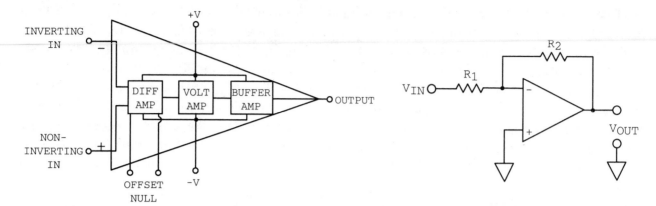

Figure 2-61. Stages of an operational amplifier (op-amp).

Figure 2-62. The inverting op-amp configuration.

Figure 2-63. The non-inverting op-amp configuration.

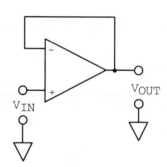

Figure 2-64. The voltage follower op-amp configuration.

Figure 2-65. A vacuum tube diode.

Figure 2-66. A vacuum tube triode.

must exist for amplification to occur. R_1 and R_2 provide voltage divider bias for Q_1, R_6 and R_7 supply voltage divider bias for Q_2, while R_4 is used as the emitter resistor for Q_1 and Q_2. R_3 and R_5 are employed as collector load resistors. R4 is typically replaced with a constant current source for improved *common-mode rejection* (CMMR) by allowing only a set amount of current to flow in the circuit, which will not permit a common-mode signal to increase this total circuit current. A *single-ended input* configuration is possible (*Figure 2-60*) in which one input is grounded with the other being used for the input signal, while the output is taken between V_1 and V_2. Other less common configurations, such as placing two out-of-phase (push-pull) signals at the input and taking the output from between V_1 and V_2 (*differential input*), or taking the output from either V_1 or V_2 to ground (*single-ended output*) are sometimes encountered.

2-8C. *Operational Amplifiers*

IC *op-amps* are frequently utilized for DC and low frequency AC amplification, while a few new types are designed to operate in the low RF range. Op-amps have very high gain, very high input impedance, and very low output impedance. Without feedback all op-amps have a very narrow BW but a very high voltage gain. To provide the operational amplifier with both a negative and positive voltage swing, a positive-negative power supply is used to power the IC. The input stage of an op-amp is a differential amplifier, followed by high gain voltage amplifiers and a high output current buffer amplifier (*Figure 2-61*). Open loop gain (no negative feedback) can be very high, but closed loop configurations, used to supply the degenerative feedback, are normally employed. The feedback, while lowering the gain substantially, increases the op-amp's bandwidth while lowering distortion and eliminating the possibility of the amplifier breaking out into oscillation. Depending on use, the op-amp is commonly configured for amplification in three ways.

First, the *inverting op-amp* (*Figure 2-62*) configuration has a gain as set by the ratio between R_2 and R_1, or $A = R_2 / R_1$. R_2 supplies the negative feedback. The lower the value of R_2 the more feedback and the lower the gain but, as stated above, the wider the bandwidth. The entire input signal is dropped across R_1, with the input impedance essentially being the value of R_1 ($Z_{IN} = R_1$). In this configuration R_1 is usually of a very high resistance, giving it a high to medium Z_{IN}, while the output impedance is usually very low. Because the signal is placed at the inverting input, there is a 180° phase shift between the input and output signal.

Secondly, with *non-inverting configurations* (*Figure 2-63*), the input is placed at the non-inverting input, with the feedback still at the inverting input to obtain the necessary negative feedback. R_2 supplies degenerative feedback because it furnishes an inphase signal to the inverting input, which lowers the difference between the - and + inputs, consequently lowering V_{OUT} and A_V. This configuration has a very high input impedance, a very low output impedance, and no phase shift between the input and output. The gain is equal to the ratio between R_2 and R_1 plus 1, or $A = R_2 / R_1 + 1$.

The third amplifier configuration for an op-amp is the *voltage follower* (*Figure 2-64*), which is used as an emitter follower. In other words, is has an extremely high input resistance and an extremely low output resistance, with no phase shift between the input and output. The voltage follower has a voltage gain of one, so it is used as a buffer or an impedance matcher between the stages, or between a stage and its load.

Some important op-amp specifications:

AV: Voltage gain of an op-amp in open loop configuration (no degenerative feedback). Specified in volts. EX: 200V/mV (For every 1mV input, 200V output will attempt to occur, or a voltage gain of 200,000).

CMRR (Common-Mode Rejection Ratio): An op-amp's ability to reject inphase, equal amplitude signals at the inverting and non-inverting terminals.

Common-mode signal: An inphase, equal amplitude signal.

Offset-null: When an op-amp has the same potential difference between its two (differential amplifier) input terminals, the output should be zero volts. If not, a proper value resistor or trimmer can be placed between its terminals to allow the op-amp's output to be adjusted to 0V.

Slew rate: The rate that the output voltage in an op-amp can change per second without causing distortion (Ex: 1V/μs).

2-9. Vacuum Tubes

Those who feel that tubes are no longer used in modern equipment are incorrect. While low power receiving tubes are not seen today, high power triodes are employed for some power audio and RF applications, as well as power pentodes and beam power tubes, outfitted with extra large or finned plates for high heat dissipation. Vacuum tubes are used in many transmitter finals that must dissipate over several hundred watts output, not to mention most high power microwave transmitters and all older equipment still in operation.

Tubes are normally ON devices and can be considered as basically a vacuum tube diode with a *control grid*. A vacuum tube diode (*Figure 2-65*), like the semiconductor diode, conducts if biased in the forward direction (with the anode, or plate, positive in respect to the cathode) and, unlike the semiconductor version, the cathode must also be heated to the proper temperature. This heating surrounds the cathode with a space charge of electrons (*thermionic emissions*), which are now free to travel towards a positive potential. If the anode becomes negative in respect to the cathode, or heating of the cathode is stopped, then conduction will cease. The cathode can be heated indirectly by a heater element, or directly through a heated-wire cathode, using either AC or DC power.

In a *triode* (*Figure 2-66*), or three element tube, a control grid is added to the diode. When the control grid is at the same potential as the cathode, the grid does not affect cathode-to-plate current flow, which would now be at maximum. When the control grid, constructed of fine wires, is supplied with a negative potential (in respect to the cathode), the attraction of the electrons for the plate is reduced, thus lowering the plate current (*Figure 2-67*). As the control grid is made more negative, a point is reached where there is zero cathode-to-plate current flow. This is referred to as *cutoff*. For class A operation, the tube is biased somewhere between these two opposite extremes of maximum current (saturation) and cutoff (zero current). Now, any incoming signal, injected at this control grid, will add-to or subtract-from this negative DC bias, which then allows the tube to conduct more or less current through a suitable resistance, creating an amplified reproduction of the input signal.

This control grid should never go positive (unless biased by *grid-leak*), or current will flow in through the grid, possibly destroying it. Also, if the DC bias is lost to the grid, high currents will flow through the tube, likewise destroying the entire tube. A cathode resistor is sometimes placed in series with the cathode to limit this possible excessive current flow.

Vacuum tubes are voltage controlled, thus having a very high input impedance. This means they draw almost no current from the prior stage.

2-9A. *Tube Types*

There are different types of tubes for various applications, usually dependent on frequency, cost, power, noise, and other design considerations.

The triode, which was described above, is now used mainly at audio frequencies due to the large interelectrode capacitance between all the elements, causing undesirable feedback and the resultant instability and decreased performance with higher frequencies. Low frequency high powered RF amplifiers may still employ triodes, with *neutralization* used to help extend the limited frequency range. Although plate current is controlled by the control grid, changes in plate voltage also significantly affect the plate current (See *Figure 2-67*). Triodes are known for their low tube noise, which can be a huge asset in audio and RF applications.

The problem of excessive interelectrode capacitance was solved by the design of the *tetrode* (*Figure 2-68*). The interelectrode capacitance is reduced significantly by the inclusion of a fourth element, the *screen grid*, between the control grid and the plate. The extra element performs as an electrostatic shield, reducing the interelectrode capacitance by three hundred times or more. This also results in a potential higher gain than a triode. As well, the screen grid makes the plate current relatively independent of changes in plate voltage. The tetrode is now rarely used, except in a few high power RF amplifiers, because of secondary emissions caused by the positive screen grid accelerating the electrons toward the plate to a high velocity. This increased speed causes some of the electrons to bounce off of the plate, to be picked up by the positively charged screen grid. This action effectively reduces plate current, which is highly undesirable.

Neutralization may still be needed for tetrodes at high frequencies, even with the minor interelectrode capacitance present, due to the tube's high power sensitivity.

The *pentode* (*Figure 2-69*) has higher gain than a tetrode, with no neutralization typically needed at lower frequencies by virtue of the inclusion of the tetrode's screen grid. The pentode also solves the problem of secondary emissions by including a fifth element, the suppressor grid, between the screen grid and plate. The *suppressor grid* is usually placed at cathode potential (in some pentodes, the sup-

Figure 2-67. *Characteristic curves for a triode tube.*

Figure 2-68. *A vacuum tube tetrode.*

Figure 2-69. *A vacuum tube pentode.*

Figure 2-70. *Characteristic curves for a pentode tube.*

Figure 2-71. *A final tube circuit biased by grid-leak.*

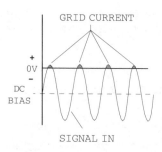

Figure 2-72. *How the input signal supplies grid current for the grid-leak capacitor.*

pressor grid is internally connected directly to the cathode). Since it is at the same potential as the cathode, any secondary emissions that occur are bounced back toward the plate to add to the plate current. A pentode's characteristic curves are shown in *Figure 2-70*. As with the tetrode, neutralization may still be needed at high frequencies, even with the minor interelectrode capacitance present, due to the tube's high power sensitivity.

Beam power tubes are high efficiency vacuum tubes with almost the same parameters as a pentode. Instead of a suppressor grid, however, it uses two beam forming electrodes at cathode potential, as well as control and screen grids that are aligned with each other. This setup repulses any secondary emissions back toward the plate and focuses the electrons into high density beams, creating a highly efficient power tube.

2-9B. Tube Amplifier Biasing

There are three major methods of biasing a vacuum tube amplifier for class A, AB, B, or C operation. The first, *grid-leak bias*, is a way to reliably and simply obtain any class of bias in an RF final tube circuit (*Figure 2-71*). The biasing components for grid-leak consists of R_1 and C_C. The class of bias depends on the amplitude of the input signal as well as the values of R_1 and C_C. Typically, however, grid-leak bias is utilized in class C power output amplifiers, since grid-leak bias has a proclivity to produce a certain amount of extra distortion of the output signal — caused when the grid conducts current to supply the needed bias. This DC grid-leak bias is created by the incoming signal going positive at the grid, causing grid current to flow (*Figure 2-72*), which charges the C_C to a negative potential on the grid, supplying the Q-point bias for the tube. With time, the charge across C_C leaks off through R_1 — with this time period dependent on the RC time constant of the two components. The bias is self-adjusting for different signal input levels, since the bias level is dependent on the signal's amplitude and frequency. The signal is clamped at a particular positive peak input voltage level, usually at about +1V at the grid. Grid current must flow in order to keep the charge on the grid-leak capacitor, and thus the necessary negative DC bias. If the signal decreases in amplitude, the capacitor will discharge through the resistor, lowering the bias voltage. Grid current then begins to flow as the top of the signal is naturally kept at +1V at the grid, which charges the capacitor to this new signal level. If the signal amplitude increases again, the negative bias, supplied by the charging capacitor, will also increase, keeping the positive peak of the incoming signal always at +1V at the grid. A resistor (R_K) is sometimes placed in series with the cathode to limit current in case all bias is lost, which would normally destroy the tube. All classes of operation from class A to C can be obtained by simply increasing the input signal level and adjusting the RC time constant of the grid-leak components.

Another form of bias, referred to as *cathode-bias* (*Figure 2-73*), is also used in low and medium powered Class A tube amplifiers. As current flows through the cathode resistor (R_K), the top of the resistor becomes more positive than the bottom and, since no current flows through the high impedance grid circuit, the cathode becomes more positive than the control grid. This is the same as making the control grid more negative, thus setting its quiescent (Q) operating point. The exact class A Q-point is set by choosing a certain value of R_K. The signal is developed across R_1, which also puts the grid at ground potential. R_K must typically be bypassed by C_K so that the constantly alternating voltage across R_K does

Figure 2-73. Cathode-bias for a tube amplifier.

Figure 2-74. A two-stage RC amplifier showing undesired shunt capacitance and frequency compensating components.

Figure 2-75. A transistor's equivalent construction at radio frequencies.

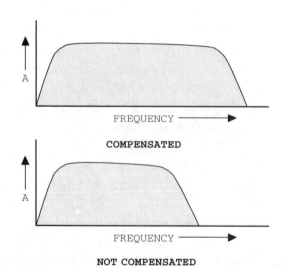

COMPENSATED

NOT COMPENSATED

Figure 2-76. The increase in upper frequency bandwidth possible with frequency compensation.

not change the bias along with the constant change in the alternating cathode current, which would substantially lower gain through degenerative feedback.

Fixed-bias can be used to set an amplifier in class A, AB, B, or C operation, and uses a power supply or supplies with separate bias voltages for the control grid and plate.

2-10. Amplifier Frequency Limitations and Compensation

All amplifiers have an upper frequency limit dictated by the internal composition and physical structure of the semiconductor or vacuum tube, as well as its bias and coupling components and its configuration. There are, however, techniques that are employed to extend these frequency limitations.

2-10A. Frequency Limitations

Transit time problems, as well as other conditions in an amplifier, limit an amplifier's ability to increase the gain of a signal at higher frequencies. Transit time refers to the action caused by a frequency being so high that it begins to take longer for an electron to pass through the base of a transistor than it does for the frequency of the input signal to change polarity, which consequently builds up an excess of electrons in the base. This action drastically lowers gain. As well, *shunt capacitance* causes lowered frequency response. *Figure 2-74* demonstrates the shunt capacitance of a standard two-stage RC coupled amplifier, while *Figure 2-75* shows a transistor's internal equivalent structure at high frequencies. C_S is the stray wiring capacitance, and shorts some of the high frequency signal to ground, as does the collector-to-emitter capacitance (C_{CE}) and the base-to-emitter capacitance (C_{BE}). *Miller effect capacitance* (C_M) is the varying (gain dependent) collector-to-base capacitance that is $180°$ out of phase with the base input, causing a degenerative AC feedback from collector-to-base at high frequencies, thus lowering V_{IN} and, consequently, gain.

f_{co} is the high frequency cutoff when gain in an amplifier drops to 70.7% of its low frequency value. This f_{co} can be extended, at the expense of overall gain, when a lower value of load resistor is used in the amplifier because $f_{co} = 1/2\pi R_L C_T$. The symbol used in the data books to indicate when a CE configured transistor's β in a test circuit reaches 0.707 of its low-frequency value is referred to as f_{ae}, $f\beta$, or f_{hfe}. Another prevalent parameter for common-emitter circuits is referred to as ft, and is the frequency at which the current gain reaches a value of unity, or 1.

2-10B. Frequency Compensation

To externally compensate for a low f_{co} in a CE amplifier (See *Figure 2-74*), *peaking coils* can be utilized in the amplifier circuit by using a coil in *shunt* (L_{P1}), *series* (L_{P2}), or in *combination*, called *series-shunt*. The shunt peaking coil L_{P1} almost cancels C_T (the total input and output shunt capacitances), which consists essentially of C_{CE} (collector-to-emitter capacitance), C_S (stray wiring capacitance), C_M (Miller capacitance) and C_{BE} (base-to-emitter capacitance) to ground. The shunt peaking coil is in shunt, or parallel, with these capacitances. L_{P1} increases the amplifier's frequency response by resonating with C_T, causing an increase in output impedance at higher frequencies, heightening stage gain. When employing a series peaking coil (L_{P2}), the output capacitance of the first stage and the input capacitance of the second stage are effectively detached and separated, allowing for a higher value of load resistor, which increases overall gain. The Q of L_{P2} is usually lowered by a parallel "swamping" resistor (R_S) to prevent coil resonance, which can overcompensate for a certain range of frequencies, causing *ringing*. When both L_{P1} and L_{P2} are used in combination, L_{P1} offsets the output capacitance of the first stage, while L_{P2} is used to counterbalance the input capacitance of the second stage. This combination can dramatically increase the frequency response across a very wide bandwidth at the output of the two coupled stages. (*Figure 2-76.*)

A designer can also take into account the frequency responses of the three transistor circuit configurations (the common-emitter, the common-collector, and the common-base) to increase this upper frequency limit. The common-base and common-collector configurations have a higher frequency response as compared to the common-emitter configuration. Unfortunately, the CB has a current gain that is always less than unity, while the CC always has a voltage gain that is also less than unity. By joining the CE and CC, or CE and CB amplifiers, however, increased frequency response and gain can be obtained.

Internal compensation at the time of manufacture can significantly increase the frequency response of a semiconductor. By making the base region as thin as possible, and by manufacturing the transistor with gallium arsenide, the transit time of an electron can be considerably shortened. NPN transistors also have faster transit times than a comparable PNP transistor due to the more rapid movement of electrons as majority carriers than that of holes through semiconductor materials.

CHAPTER 3
Oscillators
and Frequency Synthesizers

There are many ways to generate both low and high frequency sinusoidal waves. In the past, the exclusive use of RC, LC, and crystal oscillators performed this function. Now, a newer technique is being used to supply both high and low frequencies over a set design range. This method is referred to as *frequency synthesis*. Oscillators are still, of course, extensively used, and come in many different configurations, depending on the desired frequency stability, range, adjustability, and cost.

Frequency synthesizers come in two main forms: the *phase-locked loop* and a newer method called *direct digital synthesis* (DDS).

3-1. Sine-Wave Oscillators

Since all tank circuits will ring (that is, produce a decaying sinusoidal wave, *Figure 3-1*) at their resonant frequency when a pulse is applied to them, an oscillator can use an active device, such as a transistor, to amplify and sustain these pulses in the frequency-determining tank circuit. This tank circuit's natural resonant frequency is determined by the tank's L and C components, and can be calculated by using the formula:

$$fr = \frac{1}{2\pi \sqrt{LC}}$$

Oscillators use a small part of their output to feed a positive, or regenerative, feedback into their own inputs to create these oscillations. In other words, the transistor constantly amplifies its own feedback. These oscillations begin by noise pulses during turn-on, after which an oscillator is basically converting DC power into AC oscillations. Since the output of most oscillators is 180° out of phase with the input (*Figure 3-2*), some sort of phase-shifting is needed to supply the needed inphase, or regenerative, signal to its input (*Figure 3-3*). The reactance of inductors and capacitors perform this necessary phase shift in an LC oscillator.

Both the common-emitter and the common-base configuration can be used as an oscillator, with the common-emitter being the most common. The common-base oscillator is used when higher frequencies must be generated than can be handled by the common-emitter circuit. Even though there is no phase shift between a common-base's input and output, phase shifting networks are still required, however, due to a slight phase shift (60° or less) introduced by the transistor when operated at frequencies above its *alpha cutoff* (when the gain of the transistor reaches 0.707 of its low-frequency value in its CB configuration).

Figure 3-1. *Ringing of a resonant tank after application of a pulse.*

Figure 3-2. *An amplifier without a phase-shifting network generally supplies degenerative feedback — no oscillations possible.*

Figure 3-3. *Amplifier with a phase-shifting network to supply regenerative feedback — oscillations possible.*

Figure 3-4. *A UHF oscillator sometimes uses a virtual tank formed by wire inductances and capacitances for phase-shifting.*

Figure 3-5. *An LC Hartley RF oscillator. Note the tapped coil.*

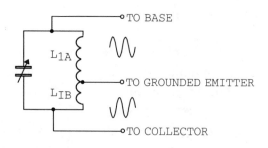

Figure 3-6. *A Hartley tapped coil with phase relationships.*

Most oscillators are biased at class C, though some are biased at class A or AB, and typically have output powers of one watt or less.

Oscillators in the UHF region sometimes take advantage of the collector-to-base capacitance, and the resultant positive feedback, to create an oscillator without a physical phase-inversion feedback tank (*Figure 3-4*). Wire inductances and capacitances at high frequencies create a "virtual" resonant tank, causing a phase shift and regeneration, which produces oscillation.

There are three main sine-wave oscillator classifications: the LC oscillator, the crystal oscillator, and the RC oscillator.

3-1A. *LC Oscillators*

LC oscillators are generally *variable frequency oscillators* (VFOs) because they can readily be tuned to set the frequency of oscillation over a wide range by adjusting either the capacitance or, less frequently, the inductance, in the LC tank. For any LC oscillator to be even reasonably stable it needs a good C to L ratio (for a high Q); a steady and separate power supply; constant temperature conditions; and oscillator isolation from its load. However, they still drift in frequency by 1% or more due to inherent real-world instability, which occurs when the temperature varies, the components age, or even when a hand is placed near the oscillator circuit (*hand-capacitance*). This is unacceptable for most modern transmitter and receiver uses without some form of frequency regulation (AFC, or *automatic frequency control*).

A type of LC oscillator called a *voltage-controlled oscillator* (VCO), however, is quite common, and is utilized in frequency synthesis schemes, such as PLLs (*phase-locked loops*), and in any application were a DC control voltage is needed to change the frequency of an oscillator.

Both the *Hartley* and the *Colpitts* oscillators are the most common of the many LC designs that may be encountered.

The Hartley (*Figure 3-5*) uses a tapped coil in its tank circuit, comprised of L_1 and C_1, to invert the phase of the feedback signal to supply a positive voltage to the base through the coupling capacitor C_2, as well as set its frequency of operation. L_2 decouples the oscillator output from being injected into the power supply and also acts as a collector load. R_1 and R_2 supply the forward bias for the transistor, while C_3 allows only AC to pass to the tank, which increases oscillator stability and lowers power losses by not allowing DC to flow in the inductor. R_E and C_4 further (temperature) stabilize the circuit as well as supply a certain amount of self-bias. The Hartley's tapped coil works because the signal between the center tap and the top of the coil (*Figure 3-6*) is opposite in polarity than between the center tap and the bottom of the coil, due to current flow with respect to the grounded tap. This furnishes the necessary 180° phase shift needed for regenerative feedback, and thus oscillations. The amplitude of the positive feedback is controlled by the location of the tap on the coil.

Figure 3-7. *An LC Colpitts RF oscillator. Note the tapped double capacitors.*

Figure 3-8. *A Hartley based voltage-controlled oscillator (VCO).*

Figure 3-9. *A Hartley crystal RF oscillator.*

Figure 3-10. *A Colpitts crystal RF oscillator.*

Figure 3-11. *A Pierce crystal oscillator.*

Figure 3-12. *Transistorized TPTG crystal oscillator.*

The Colpitts (*Figure 3-7*) LC oscillator is essentially the same as the Hartley, but instead of a tapped coil sent to ground in the tank circuit, it uses twin capacitors tapped to ground. This has the same effect as the tapped coil by creating a 180° phase inversion across each capacitor, and supplying the positive feedback into the base as needed for oscillation. Since the capacitance ratio between the two capacitors changes the feedback voltage, the capacitors are always ganged to change the oscillator's frequency. A tunable inductor can also be used, with the capacitors placed at fixed values.

As stated above, VCOs are LC oscillators that can vary their output frequency by the application of a DC control voltage across *varactor diodes* (a voltage-controlled capacitor), which then varies its capacitance in the tank of an LC oscillator, changing its frequency either above or below a *rest* or *center frequency*. This rest frequency is set by the DC bias across the varactors, placing their capacitance at a fixed and unvarying value. By adding to or subtracting from this bias, the frequency of the oscillator can be changed. In the example of *Figure 3-8*, Q_1 and its associated components function as a Hartley oscillator, while Q_2 performs as a buffer for the VCO's output (so that the oscillator is not loaded down by the low input impedance of the next stage). It is commonly used in PLLs, AFCs, and older manual electronic tuning circuits.

3-1B. Crystal Oscillators

It is obvious that modern equipment cannot function properly with the excessive drift of even the best of the LC oscillators. Not only would a transmitter drift and interfere with adjacent channels, but its received signal would be unreadable as it drifted into and out of the passband of the receiver, increasing and decreasing in volume, pitch, and/or distortion. The receiver itself, if not frequency stabilized, would be hopelessly inadequate for most uses, especially in FM and SSB applications, as most signals would be hopelessly unintelligible due to excessive LO drift. One way to create a virtually drift-free oscillator output is to utilize some type of piezoelectric crystal. A crystal will produce a voltage if mechanical stresses are placed upon it, and will likewise produce mechanical expansion and contraction when an external voltage is placed across the crystal. In other words, if an alternating signal of the proper frequency is placed across the crystal, it will vibrate at its natural resonant frequency only. There are many types of crystals, but the most common is quartz because of its physical strength, heat stability, and low cost. It is the crystal material most commonly found as the frequency controlling element in a crystal oscillator. This ability of the crystal to only pass a signal at the crystal's natural resonant frequency makes it perfect for its application in oscillators.

Crystal oscillators cannot readily change frequency to any appreciable degree, but may be tuned by a few hertz with a capacitive trimmer in parallel or series with the crystal. Rarely, and only in low power applications, a crystal's frequency can be varied slightly by using an adjustable air-gap between one plate, or both plates, and the face(s) of the crystal. This type of tuning works because the spacing between the plates and the crystal, as well as the plate's pressure upon the crystal, affects its frequency.

Maximum crystal frequency can range up to 50 MHz on its fundamental frequency, yet it can reach higher frequencies on harmonic operation. A *harmonic crystal oscillator* can have an output of its second, third or fourth harmonic, with the tank circuits tuned to output one of these harmonics. How-

Figure 3-13. An audio RC phase-shift oscillator.

Figure 3-14. An audio RC Wien-bridge oscillator.

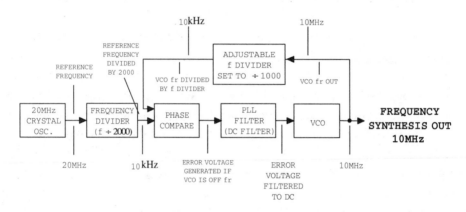

Figure 3-15. The basic sections of a phase-locked loop frequency synthesizer.

Figure 3-16. A premixing circuit employing a PLL and a crystal oscillator.

Figure 3-17. One scheme using direct digital synthesis (DDS) to synthesize new frequencies.

ever, the crystal itself is actually operating at its lower fundamental resonant frequency. An *overtone crystal oscillator*, because the crystal must actually vibrate at high harmonic frequencies, will function strictly at its third or fifth harmonic — with the output tank tuned to this chosen output frequency. A special overtone crystal is normally used for this type of VHF operation, and will only oscillate at an odd harmonic of its fundamental.

There are many distinct classifications of crystal oscillators; therefore, we will only concentrate on the most common.

With the *Hartley crystal oscillator* (*Figure 3-9*) and the *Colpitts crystal oscillator* (*Figure 3-10*) the crystal is placed in the feedback path. Since a crystal has a very high Q (in excess of 50,000) and thus a very narrow BW (*BW = fr/Q*), the series crystal will only allow the feedback current at its resonant frequency to pass onto the phase shifting LC circuits and into the oscillator's input. Any feedback that is even slightly off frequency will be severely attenuated, thus lowering feedback, and forcing the oscillator back on frequency. This makes the normal LC Hartley and Colpitts oscillators "rock steady".

The *Pierce crystal oscillator* (*Figure 3-11*) employs a different principle of operation. The inductor is replaced by a crystal operating in its slightly higher parallel resonant mode, which to the tank circuit appears electrically as an inductor. (Any crystal placed in its holder also has a higher *parallel* resonant frequency, as well as the lower series resonant frequency, caused by holding plate capacitance). Since impedance is at maximum at this parallel resonant frequency, the feedback across C_1, and thus into the base, is also at maximum, forcing the oscillator on frequency. The Pierce oscillator is one of the most popular crystal oscillators for use in low power, medium frequency applications, and possesses a high frequency stability.

Another popular crystal oscillator is sometimes referred to as a TPTG (tuned-plate tuned-grid, *Figure 3-12*) crystal oscillator. The name is taken from an old vacuum tube LC oscillator that used twin resonant tanks; one at the input, the other at the output. In the semiconductor TPTG oscillator, there are two tuned circuits, one in the base, consisting essentially of the crystal, and the tuned circuit at the collector, consisting of an LC output tank. R_1 and R_2 set the transistor's bias, L furnishes an elevated impedance for the crystal output into the base, while R_E provides temperature stabilization and a certain amount of self-bias, with C_1 bypassing RF around R_E for increased AC gain.

3-1C. RC Oscillators

Two types of sine-wave oscillators are commonly used as audio frequency generators: the *phase-shift* and the *Wein-bridge* oscillator. Both employ resistor-capacitor (RC) networks as the frequency setting components, inasmuch as the utilization of LC components would dictate the use of large and expensive parts at these very low frequencies.

The *phase-shift oscillator* (*Figure 3-13*) uses the phase shifting abilities of an RC network to create the necessary positive feedback at the input to the oscillator. Since in a totally capacitive circuit the current leads the voltage by 90°, combining a capacitor with a resistor (with its zero phase shift), creates some phase shift value between 0 and 90°, depending on each component's values. Typically three 60° phase shifts can be employed, using three pairs of resistor-capacitor components to supply the full 180° phase shift needed for positive feedback. Since capacitors vary in reactance depending on the frequency of the alternating current, any frequency that does not give the proper reactance to the capacitors necessary to supply the combined 180° RC phase shift, and its resultant positive feedback, will not reach the oscillator's input, and will consequently force the oscillator back on frequency to a frequency that *does* allow an exact 180° phase shift between its output and input. This will supply the input to the oscillator with a regenerative (positive) feedback, and thus sustain continued oscillations at the proper design frequency. The phase-shift oscillator is usually employed only in fixed frequency applications.

The second type of common low frequency oscillator is the *Wien-bridge* (*Figure 3-14*), which can be used in variable-frequency operations to output high quality sinusoidal waves. The RC network, called a *lead-lag network*, sets the operating frequency of the oscillator by supplying both a regenerative and degenerative feedback, and ordinarily is operated in conjunction with an IC op-amp rather than any discrete amplifiers. As long as the feedback is slightly positive, oscillations will occur, but at any value where it becomes negative, no oscillations result, forcing the oscillator back on frequency. Actual phase shifting of the feedback voltage does not take place — the network merely chooses the frequency at which this feedback occurs. R_1 - C_1 and R_2 - C_2 furnish the regenerative feedback to the positive, or non-inverting, input to the op-amp, while the degenerative feedback is created across R_3 and R_4 and placed at the negative, or inverting, input. At the oscillator's resonant frequency, the regenerative feedback across the R_1 - C_1 and R_2 - C_2 combination exceeds the degenerative feedback voltage present across R_3 - R_4, forcing the oscillator to stay on frequency. C_1 and C_2 can be ganged to vary the output frequency.

3-2. Frequency Synthesis

Frequency synthesis is a technique for generating extremely accurate multiple frequencies from one (but frequently more) crystals. It has become the dominant method for variable frequency production in receivers, transmitters, transceivers, and test equipment. There are many ways to obtain frequency synthesis, but the most common today is by the *phase-locked loop* (PLL) and a newer system called *direct digital synthesis* (DDS). Both methods can vary tremendously in circuit layout from one PLL circuit to the next, or from one DDS system to the next, depending on the frequency accuracy, stability, the number of simultaneous frequencies needed, upper frequency limits, noise generation, tuning resolution, and the bandwidth of tuning desired. Obviously then, only the basics of frequency synthesis can be covered here.

3-2A. Phase-Locked Loop

Most modern frequency synthesis is based on the phase-locked loop (PLL). *Figure 3-15* shows a simple PLL frequency synthesizer. A crystal oscillator frequency (the *reference frequency*) is fed into a fixed *frequency divider*, which lowers the reference frequency by a set amount and places it into the *phase-comparator*, which compares the frequency from the fixed frequency divider to that of the *adjustable frequency divider*. The adjustable frequency divider acquires its frequency from the output of the VCO (*voltage-controlled oscillator*, or "voltage-to-frequency converter"). The phase-comparator compares the two frequencies, and if they are the same, no error voltage is outputted from the phase-comparator. If they differ, then a rectified DC error voltage is outputted to the *PLL filter*, which removes any AC variations and noise, and feeds this DC control voltage directly into the VCO (or, in some PLLs, first into a DC amplifier and then into the VCO). A varactor's bias in the VCO is affected by this DC control voltage, which forces the VCO back on frequency if it has drifted off. The adjustable frequency divider is now commonly controlled by the operator through a panel knob, with a microprocessor supplying digital control words (by a serial or parallel bus). The microprocessor can also be used to decode and drive the frequency display circuits (either LEDs or LCDs), so that the operator will be visually updated on any frequency changes that may be made. Many IC PLLs are designed to output two frequencies for frequency synthesis: one frequency as the transmitter's master oscillator, and another for input into the transmit mixer; or for a dual conversion receiver's two LO frequencies. Many PLL communications gear may have more than one PLL chip and/or utilize external VCOs, as well as employ simple mechanical push-buttons or switches for channel changes into the PLL chip inputs.

For PLL circuits capable of frequency synthesis at higher frequencies, the output of a crystal oscillator and the PLL can be fed into a mixer, filtered to obtain the sum or difference of the two frequencies, resulting in a higher synthesized frequency. This is referred to as *pre-mixing* (*Figure 3-16*).

3-2B. *Direct Digital Synthesis*

A very recent development in frequency synthesis is called DDS (Direct Digital Synthesis). DDS lowers the cost and intricacy of most common frequency synthesizers with a much improved frequency resolution, giving the feel of analog fine tuning with up to 1 Hz resolution (tuning resolution is the fineness of the frequency steps the tuner is capable of — the lower the better). DDS usually comes in small surface mount ICs, along with one or more support ICs. DDS works by utilizing a crystal oscillator as a reference frequency, digitally calculating the difference between this reference and that of the frequency the radio operator wishes to generate. This new calculated digital signal is fed to a D/A converter and then placed into a low-pass filter (LPF) — which outputs the new synthetic digitally constructed sinewave. A pure DDS system has limited frequency range, so DDS is usually combined with standard PLLs to increase the frequency to usable levels. Also, since it is a pure digital form of frequency synthesis, a type of noise is created called quantization noise. This simply refers to errors made when the DDS output is converted from digital to analog. The analog synthesized frequency, with its infinite amount of possible amplitudes, is actually being constructed from a multitude (but finite number) of discrete digital levels — making complete accuracy during conversion impossible.

One way to implement a practical RF DDS frequency synthesis scheme is as follows — when the operator changes frequencies, the radio's on-board microprocessor outputs control words into the DDS (*Figure 3-17*) for a frequency change, which then synthesizes a low, stable synthetic analog frequency that is fed into a summing loop (a complex circuit that performs a type of premixing). The loop adds the low frequency output of the DDS to a high frequency reference, mixing the subsequent output into a much higher stable frequency. Since the frequency variations possible at the output of the summing loop are quite small, they are then inputted into a PLL (with its own crystal reference), along with coarse frequency digital control words. The output is now the desired wide-ranging, but rock steady, synthesized frequency. There are many other ways to achieve a functional DDS frequency synthesis system.

CHAPTER 4
Amplitude Modulation

Amplitude modulation (AM) is one of the earliest forms of wireless communications. It is extremely simple and low in cost to implement. Due to this low cost and simplicity, it is still extensively used — mainly in commercial and shortwave broadcast and citizens-band radio operation. Understanding AM is vital for the comprehension of newer means of radio communications.

4-1. AM Fundamentals

Modulation is the process of impressing information on a carrier wave, be it voice, digital code, images, etc. Demodulation is the process of removing this information and sending it to any number of transducers (such as a speaker or headphones for voice and music), or to a cathode ray tube (CRT) for visual display, or through to digital circuits for processing or storage.

4-1A. Modulation and Demodulation

The simplest way of impressing voice or music on a carrier is by amplitude modulation. The unmodulated carrier (*Figure 4-2a*) is the radio frequency signal and transports any modulation, through space, to the receiver. The modulation is the *intelligence* (*Figure 4-1b*), always at a lower frequency than that of the carrier, which is impressed on this carrier through nonlinear mixing. The *amplitude* of the carrier is altered, when viewed in the time-domain, at the rate of the modulating signal's *amplitude* and *frequency variations*. Increasing the amplitude of the modulating signal increases the amplitude of the carrier (*Figures 4-1c & 4-2c*), while decreasing the modulation's amplitude decreases the carrier's amplitude. (*Figures 4-1b & 4-2b.*) The modulation "rides" the carrier to the receiver, where these amplitude variations are stripped off of the carrier and converted into audio amplitude variations (*Figures 4-3b & 4-3c*) and amplified, then fed to an appropriate transducer. The percent of modulation (*Figures 4-2b & 4-2c*) will govern the amplitude of the final detected signal (*Figures 4-3b & 4-3c*) and the resultant audio volume at the receiver's speaker.

In class C collector modulation (or plate modulation, in vacuum tube finals), both the positive and negative alternations of the carrier are affected equally. The negative alternation is recreated by the output tank of the transmitter's final amplifier, generating a mirror image of the positive modulated alternation (*Figure 4-4*).

Modulation between a carrier and its baseband signal (voice, music, code, etc.) creates sidebands, viewable in the frequency-domain (*Figure 4-5*), due to the nonlinear mixing creating sum (USB, or *upper sideband*) and difference (LSB, or *lower sideband*) frequencies. The phase relationships between the carrier and the LSB and USB cause a resultant new waveform that varies in amplitude (*Figure 4-6*). When the sidebands and carrier are in phase, the amplitude of this new waveform is twice that of the

INTELLIGENCE (AUDIO)	CARRIER (AFTER MODULATION)	AFTER DEMODULATION

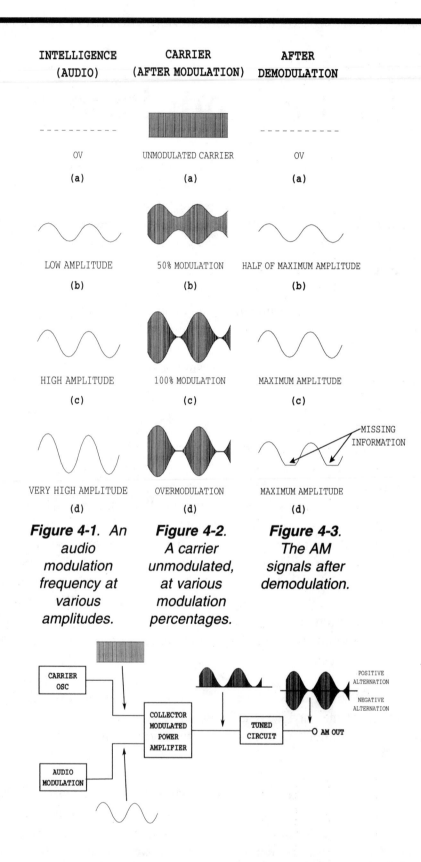

OV

UNMODULATED CARRIER

OV

(a)   (a)   (a)

LOW AMPLITUDE   50% MODULATION   HALF OF MAXIMUM AMPLITUDE

(b)   (b)   (b)

HIGH AMPLITUDE   100% MODULATION   MAXIMUM AMPLITUDE

(c)   (c)   (c)

MISSING INFORMATION

VERY HIGH AMPLITUDE   OVERMODULATION   MAXIMUM AMPLITUDE

(d)   (d)   (d)

Figure 4-1. An audio modulation frequency at various amplitudes.

Figure 4-2. A carrier unmodulated, at various modulation percentages.

Figure 4-3. The AM signals after demodulation.

CARRIER OSC

COLLECTOR MODULATED POWER AMPLIFIER

TUNED CIRCUIT

AM OUT

POSITIVE ALTERNATION

NEGATIVE ALTERNATION

AUDIO MODULATION

Figure 4-4. AM generation in a class C collector modulator.

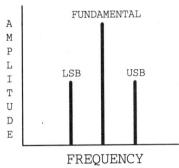

Figure 4-5. An AM carrier and its sidebands when modulated by a single tone.

Figure 4-6. A single tone amplitude modulated carrier in the time-domain.

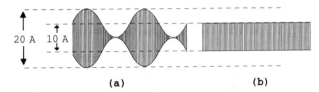

Figure 4-7. A 100% amplitude modulated carrier (a); with unmodulated carrier (b).

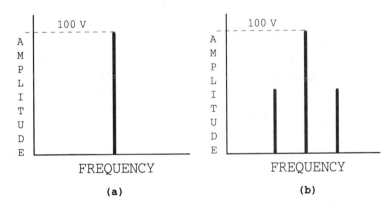

Figure 4-8. The carrier amplitude of an AM signal remains unchanged, regardless of modulation: (a) unmodulated carrier; (b) modulated carrier.

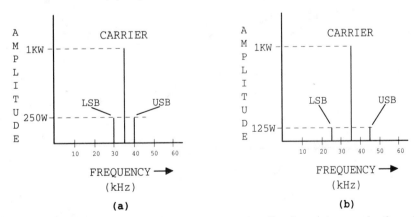

Figure 4-9. When the modulation frequency or amplitude varies, only the sidebands are altered: (a) 5 kHz high amplitude modulation; (b) 10 kHz low amplitude modulation.

Figure 4-10. Voice signals contain sidebands at various frequencies and amplitudes.

Figure 4-11. A basic AM superheterodyne receiver.

Figure 4-12. The single diode AM detector.

Figure 4-13. A full-wave AM diode detector.

Figure 4-14. Common placement of a receiver volume control.

Figure 4-15. Low-level collector modulation.

unmodulated carrier. When the two sidebands are completely out-of-phase with the carrier, the amplitude of the new waveform is almost nil. In other words, in the time-domain, the amplitude of this new modulated carrier waveform is twice that of the unmodulated carrier at the peaks with 100% modulation, and zero in amplitude at the troughs (*Figure 4-7*). When viewed in the frequency-domain on a spectrum analyzer, the carrier's frequency and amplitude do not vary, whether modulated or not (*Figure 4-8*). This is an important concept, because this proves that the carrier itself contains no information — the information is contained in the sidebands only. When amplitude modulation is viewed on the spectrum analyzer, it can be seen that when the modulation is changed in frequency and amplitude at the transmitter, the sidebands will change frequency and amplitude, while the carrier will remain at its original frequency and amplitude, indicating that there is no information actually contained in the carrier, but only in the sidebands (*Figure 4-9*). Each sideband will contain the exact same information and power as the other.

The percent of modulation can be found in the time-domain, on an oscilloscope display, by the formula $V_{PEAK} - V_{MIN} / V_{PEAK} + V_{MIN} x\ 100$. As mentioned, when the carrier wave is modulated by an audio wave, sidebands are produced, as can be observed only in the frequency-domain, on a spectrum analyzer display. These sidebands are at sum and difference frequencies of the carrier and modulating frequencies, or $f_{CARRIER} - f_{AUDIO} = LSB$ and $f_{CARRIER} + f_{AUDIO} = USB$. For example, if a 100 kHz carrier is modulated with a 5 kHz audio tone, the sideband frequencies are 100 kHz + 5 kHz or 105 kHz, and 100 kHz - 5kHz or 95 kHz. The bandwidth of this resulting signal is double the audio frequency ($BW = 2\ x f_{AUDIO}$ or $USB - LSB$) or, in the above example, $2\ x\ 5\ kHz$, or 10 kHz. Since the carrier power remains unchanged no matter what the modulation percentage, the total power of an AM signal equals the sum of the carrier and sideband powers, or $P_T = P_C + P_{LSB} + P_{USB}$. Only the *total* AM power varies. When the AM carrier is modulated, antenna current increases due to the power added by the sidebands.

An example of a voice amplitude modulated signal as might be displayed on a *spectrum analyzer* screen can be seen in *Figure 4-10*. Many sidebands are produced at varying frequencies and amplitudes, depending on the audio modulation's own amplitudes and frequencies.

4-1B. Overmodulation

Most AM voice transmitters limit modulation frequencies from 300 to 3000 Hz by the use of a BPF (bandpass filter), placed after the first audio (microphone) amplifier, to limit transmitted bandwidth. A *limiter* can also be used, to limit the maximum audio amplitudes into the audio modulator so that *overmodulation* cannot occur. Overmodulation causes distortion and an increase in bandwidth due to *splatter*. Splatter is harmonic generation of the original audio frequencies caused by clipping of the modulation envelope, which further modulates the carrier, causing adjacent channel interference — which is heard as a crackling noise in the receiver for many hundreds of kilohertz from the carrier frequency. Also, since part of the audio envelope is not present (see *Figures 4-2d & 4-3d*), intelligibility of the received signal is also degraded.

Figure 4-16. *Low-level emitter modulation.*

Figure 4-17. *Low-level base modulation.*

4-1C. AM Power Measurement

In AM the *peak envelope power* (PEP) can be used to measure the peak instantaneous power, with 100% modulation applied, of the transmitted signal ($PEP = V^2_{RMS}/R$ or $V_{RMS} \times I_{RMS}$ or $I^2_{RMS} \times R$ of the peak values at 100% modulation), or the carrier power can be calculated using the same formulas, with zero modulation applied.

4-2. AM Receivers

AM receivers have much in common with almost all other RF receivers, in that all typically apply the superheterodyne principle, usually varying only in the form of demodulation of the incoming modulated radio signal.

The quality of a modern AM receiver — its sensitivity, selectivity, noise figure and, to a lesser extent, its dynamic range — vary between communication sets. However, all employ basically the same circuits, in varying numbers (whether in IC or discrete forms), to convert the RF signal into an intermediate frequency, strip the audio from the IF, and amplify the resultant AF into a speaker or headphones.

4-2A. AM Superheterodyne Circuits

Most AM receivers exploit the *superheterodyne* principle. All AM superheterodyne receivers (*Figure 4-11*) receive a high frequency modulated signal (the RF), converting it to a lower single fixed intermediate frequency (the IF). Conversion is accomplished by mixing the incoming RF with a local oscillator (LO), then filtering the mixer's output to obtain the difference frequency. This difference frequency (now called the IF) is applied to high gain, narrow selectivity IF amplifiers. The IF is then sent to a demodulator, such as the simple *diode detector* of *Figure 4-12*, where the modulated IF is rectified. (The RF is changed into a pulsating DC by cutting off the IF's negative half cycles.) The positive DC pulses of the IF itself now charge up the capacitor. Between these DC pulses the capacitor is able to discharge through the resistor, but the RC time constant is such that the charge and discharge cycle mimics the audio envelope riding on the IF — and the audio envelope is thus recovered. A *full-wave AM diode detector* of *Figure 4-13* can also be utilized. The full-wave detector produces a higher and easier to filter output, as well as less distortion and attenuation of the higher audio frequencies, than the single diode type. The resultant audio frequency from any AM detector is then fed to the audio amplifiers and on to the speaker.

The audio stages of a receiver are usually class A or linear class AB or B push-pull, with the final audio power amplifier almost always push-pull. Also, any AM amplifier stage that contains a modulated signal, such as the RF and IF stages, cannot use nonlinear stages (such as class C), since the incoming signal contains a modulation envelope that would be severely distorted by the non-linearity.

The front end of a receiver consists of the tuned circuits from the antenna input up to and including the mixer and local oscillator, as well as any possible radio frequency amplifier placed before the mixer. The

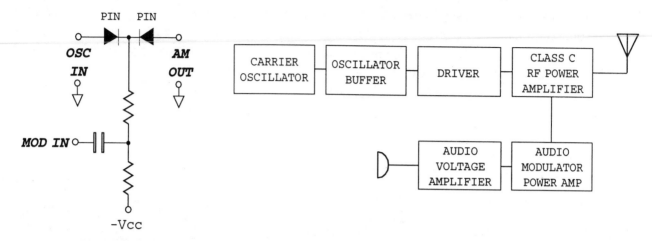

Figure 4-18. A PIN AM modulator.

Figure 4-19. AM high-level modulation.

Figure 4-20. An AM high-level collector modulator.

RF amplifier stage or stages may not be present in receivers designed for reception of frequencies under 30 MHz, but their inclusion increases sensitivity, selectivity, improves the signal-to-noise ratio, the image and spurious frequency rejection, and isolates the LO from the antenna, lowering the possibility of the LO radiating, causing interference.

The volume adjustment (audio gain) is usually located just after the audio detector and before the first audio amplifier (*Figure 4-14*).

4-2B. Image Frequency

A single conversion (one IF) AM superheterodyne receiver has a drawback in that any frequency at its chosen IF will be amplified by the IF stages and outputted into the speaker as interference. The image frequency is any frequency that is at twice the IF plus the RF signal frequency ($\{2 \cdot IF\} + RF$), or at the IF plus the LO frequency (LO + IF). In other words, it is a frequency that differs from the LO frequency by the amount of the IF, just as the desired signal does, but is higher instead of lower than the LO frequency. Subtracting the signal frequency from the LO frequency will gives us the IF (f_{IF}), which passes through the IF amplifiers. Any frequency that is *higher* than the LO frequency by the same amount that the signal frequency (fs) is below the LO frequency (f_{LO}) would give us the same frequency, and thus be passed through the IF amplifiers, causing audio interference. To make this concept a little clearer: if we had a signal frequency of 100 MHz and a LO frequency of 101 MHz, this would give us an IF of 1 MHz. Since $f_S - f_{LO} = f_{IF}$, then *100 MHz - 101 Mhz = 1 MHz*. If another *frequency* (the image frequency, f_i) happened to be at 102 MHz, with the local oscillator at 101 MHz and the IF at 1 MHz as above, then $f_i - f_{LO} = f_{IF}$, or *102 MHz - 101 MHz = 1 MHz*.

There are three ways to attenuate these image frequencies. One is to have a very high-Q preselector (an amplifier that supplies a narrow bandwidth to pass the signal frequency only) at the input of the receiver; or to have a high intermediate frequency to remove the image frequency as far away from the desired frequencies as possible; or to utilize double or triple conversion receiver designs. With multiple conversion receivers, the first IF is very high to keep the image far away from the signal, so that it will be out of the passband of the receiver. The second and third IFs are considerably lower in frequency to supply most of the selectivity and gain. (The lower the IF the more simple, stable, sensitive, and selective the amplifiers become due to the lower stray capacitances and inductances at these lower frequencies. As well, there is no need to employ special high frequency components or shielding.)

4-3. AM Transmitters

A transmitter, at its most basic, is an oscillator that can be turned on and off. This simple transmitter could be used to transmit information in the form of *Morse code*. Most transmitters, however, are far more complex. Even a Morse code (CW) transmitter ordinarily has supplemental circuits to supply added amplification for increased power output; tuning capabilities for transmission on more than one frequency; and filters to reduce harmonic and spurious radiation. Any transmitter that must transmit

Figure 4-21. *An AM vacuum tube plate modulation stage.*

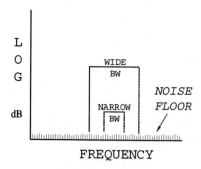

Figure 4-22. *As the receiver's bandwidth is increased, noise levels must increase.*

voice or other *baseband signals* (audio, video, or pulse data modulation) must have some method of impressing, or modulating, this low frequency onto the higher frequency RF carrier for transmission into space.

4-3A. Modulation

In AM, modulation is accomplished by two basic means. The first technique is referred to as *low-level modulation*. With this method the carrier is modulated at any stage before the final power amplifier, necessitating linear amplifiers after modulation to sustain the audio envelope without distortion. Feeding the carrier into the base and modulating the collector (*Figure 4-15*), the emitter (*Figure 4-16*), or even the base (*Figure 4-17*) can be used. *Collector modulation* is very common in high-level modulation, and will be discussed in detail below. With emitter modulation the modulating signal varies the gain of the common-emitter stage, and thus that of the carrier injected into the base. For this type of modulator to function properly, the modulation amplitude must be higher than that of the carrier. Base modulation injects both the carrier and the modulating signal into the base, with the transistor biased class C. Since class C will only pass half the waveform, the complete waveform is reconstructed by the low Q tuned output circuit. This configuration is rarely used due to its low efficiency and inferior linearity.

The main advantage of low-level modulation is that very little audio power is needed to create the amplitude modulated signal. Still, low-level modulation is seldom practiced because any energy savings enjoyed in using a low level of modulation power is offset by the need to operate with relatively inefficient linear amplifiers, whether single-ended class A, or AB or B push-pull. The exception is in low cost and low power transmitters, or in AM generation above the VHF spectrum, which typically employs *PIN diodes*. (One of the few methods to produce AM at high frequencies is with the PIN modulator, *Figure 4-18*.) The PIN diodes are forward biased by the -Vcc. This bias is subtracted from or added to by the negative and positive alternations of the modulation signal, which varies the resistance of the PIN diodes and, since these diodes are in series with the master oscillator and the load, modulates the amplitude across the "AM OUT" terminals.

The second technique employed in AM is *high-level modulation* (*Figure 4-19*), which modulates the RF carrier in the very last stage, generally using extremely efficient class C RF amplifiers. High-level modulation offers excellent linearity but, unfortunately, the audio modulator stage must be capable of supplying a peak voltage almost equal to V_{CC}, as well as half the power of the final RF power amplifier, for 100% modulation. This means that significant power must be generated in the inefficient linear audio stages.

Figure 4-20 demonstrates the typical high-level collector modulator circuit. The transistor and its bias components furnish nonlinear class C RF power amplification, while the audio signal, which is in series with the collector supply voltage through the secondary of the audio modulation transformer, either adds-to or subtracts-from the collector voltage, which will then cause either an increase or decrease in collector output current. Since this amplifier is biased at class C, only pulses are outputted across the

RFC. These pulses are then coupled into the tank network by C4, where they are changed into a complete AM waveform. (See *π Filter*, *Flywheel Effect*, and *Tank Circuit*.)

Figure 4-21 shows a high-powered tube final AM stage employing plate modulation (comparable to a transistor's collector modulation). Class C is used for this last RF amplifier because the final must be nonlinear to produce the necessary modulation sidebands that create AM. The output tank reproduces the complete carrier, with its attendant sidebands, from the pulses of this class C final amplifier, causing the output tank to oscillate at its natural resonant frequency.

The audio modulator itself would be a linear single-ended class A, or a double-ended AB or B push-pull, power audio amplifier. The modulation transformer's secondary is in series with the final's plate supply voltage, with the audio modulation adding-to or subtracting-from the plate supply, which then varies the amplitude of the output sine wave, creating AM.

To achieve solid and reliable 100% modulation with high-level modulation techniques, both the driver and the final are typically modulated, with both operating class C. Another common method modulates both the plate and the screen-grid of a pentode vacuum tube to achieve high-efficiency, reliable 100% modulation.

High-level modulation is used in all but the most powerful AM transmitters. With very high powered transmitters it becomes increasingly difficult to efficiently increase the audio power needed for high-level modulation, necessitating the use of low level modulation techniques.

4-4. AM Disadvantages

Some disadvantages of AM are as follows — most of the power is in the carrier, which is not needed to supply the intelligence. The BW is also double what it needs to be, since only one sideband is needed for transmission and reception of the intelligence. The phase relationship between the carrier and the sidebands must be exact; this is hard to maintain under many conditions. (If this phase relationship is not sustained, severe fading will result.) As an example of wasted power: with 100% modulation and a 100W carrier, the total AM power would be 150W, with 25W in each sideband — even when no speech is occurring, 100W is wasted by the carrier, with another 25W wasted when modulation occurs. Most if not all of these problems are corrected with the use of *single sideband suppressed carrier* (SSSC or SSB). SSB needs only half the bandwidth — the power conserved in not transmitting the carrier and one sideband can be placed into transmitting the single information-carrying sideband signal. Also, since there is less bandwidth with only one carrier, there is less noise received (*Figure 4-22*) by the SSB receiver. As well, far less fading occurs because no phase relationships must be maintained between the USB, LSB, and carrier. Additionally, a signal is only generated when modulation is present, since the carrier is *always* suppressed within the transmitter, further reducing wasted power.

CHAPTER 5
Frequency Modulation

FM was created to overcome the limitations of AM, which is essentially that of excessive noise. Noise is generally caused by unwanted amplitude variations, which can be removed by amplitude limiters in the FM receiver.

The method of producing FM can be accomplished by two methods: *direct FM*, which directly modifies the frequency of the FM carrier in step with the modulation's amplitude variations; and *phase modulation*, which varies the phase of the carrier, creating *indirect FM*. Both methods cause the end effect of modulating the frequency of the carrier, and are both categorized under the term *angle modulation*.

5-1. FM Fundamentals

Modulation is the process of superimposing information on a carrier wave, be it voice, music, etc. Demodulation is the process of removing this information and conveying it to any number of transducers; such as a speaker or headphones for voice and music, or through to digital circuits for processing or storage. FM performs this process of modulation by varying the frequency of the RF carrier in sync with the changing amplitudes of the modulating (baseband) signal. When the modulated carrier reaches the receiver, these baseband-created frequency variations are stripped off of the RF carrier and converted into audio amplitude variations, then amplified and fed into the appropriate transducer.

5-1A. FM Modulation

In FM the *amplitude* of the modulation signal varies the *frequency*, and not the amplitude as in AM, of the carrier (*Figure 5-1*). As the modulation signal amplitude increases, so does the amount of *frequency deviation* (the amount the carrier deviates from its center frequency in *one direction* during frequency modulation). The frequency of the carrier without modulation, or the *center frequency*, will remain at the transmitter's preset carrier frequency until modulated. When modulation occurs, the carrier increases and decreases in frequency, depending on the amplitude of the baseband signal. As the audio swings positive, the carrier increases in frequency; as the audio modulation swings negative, the frequency of the carrier swings below its rest frequency. As the *frequency* of the *modulating audio* varies, the *rate* that the frequency modulated carrier crosses the rest frequency varies at this same audio rate. For instance, if the modulating audio tone is 3 kHz, then the carrier will swing past its rest frequency three thousand times in one second.

The percent of modulation for FM is controlled by government regulation, and not by any natural limitations as it is in AM. 100% modulation in narrowband (voice) communications, for instance, is considered to be around 5 kHz deviation. For wideband (FM broadcast), 75 kHz deviation is the

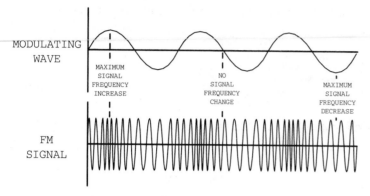

Figure 5-1. *The effect of sine wave modulation on an FM signal.*

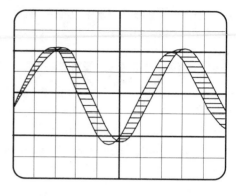

Figure 5-2. *Frequency modulation as seen in the time-domain.*

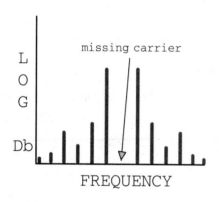

Figure 5-3. *An FM carrier null occurs at certain modulation indexes.*

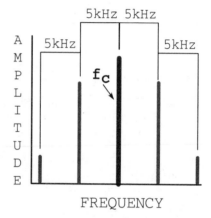

Figure 5-4. *Single-tone modulation sideband spacing from the FM carrier.*

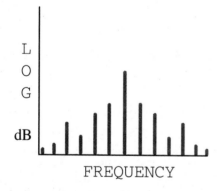

Figure 5-5. *Multiple sidebands produced by a single-tone baseband signal in FM.*

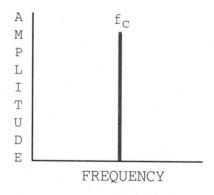

Figure 5-6. *A unmodulated FM (CW) carrier (modulation index of 0).*

maximum allowed. If the amplitude of the baseband signal causes the FM deviation to go above 100%, more sidebands are produced, which increases bandwidth, causing possible interference with adjacent channels.

When frequency modulation is viewed on an oscilloscope, the carrier can be seen not to vary in amplitude when modulated, but merely in frequency (*Figure 5-2*). This is evidenced by a shortening and lengthening of the signal's wavelength, causing a blur on the display due to the rapid frequency changes. (Since wavelength equals the speed of light divided by the frequency, or $\lambda = C/f_r$, then any change in wavelength directly corresponds to a change in frequency.)

5-1B. *Output Power and Significant Sidebands*

The combined power or voltage in a frequency modulated signal does not change when modulated or unmodulated. FM transmitter power always remains constant during modulation. Any sidebands produced must obtain all of their power from the carrier, which then must obviously lose power in the creation of these sidebands. For example, when an FM transmitter is unmodulated, a certain carrier is 100W. Therefore, when this RF carrier is modulated, the carrier must give up some (or all, depending on sideband amplitude and number) of its power to these sidebands; and the carrier and its *significant sidebands* must all add up to a total of 100W. During this modulation process, an infinite number of sidebands are generated. This is because the carrier is shifted through an infinite number of sundry frequency (or phase) values by the constantly varying modulating frequency, consequently producing an infinite number of sideband frequencies. (The amplitude of even a single test-tone frequency will, of course, vary in a sinusoidal manner with an infinite amount of amplitude variations throughout a single cycle.) Considering that "infinite" is an impossible concept both mathematically and in practice, the idea of *significant sidebands* must be used. These significant sidebands are simply any sidebands that have an amplitude that is 1% or more of the unmodulated carrier amplitude. Any sideband that is less in amplitude is not considered to be of consequence.

At certain *modulation indexes*, which is the ratio between the carrier's instantaneous frequency deviation divided by the instantaneous frequency of the modulation (or FREQUENCY DEVIATION/AUDIO MOD FREQUENCY), the sidebands will contain all of the energy, and the carrier will disappear completely (*carrier null*, *Figure 5-3*). The greater the amplitude and/or the lower the frequency of the modulation, the larger the number of significant sideband frequencies generated. The sidebands are spaced on each side of the carrier, and from themselves, by an amount equal to the frequency of the single-tone modulation signal, as shown in *Figure 5-4*. But, unlike AM, more than one pair of sidebands are produced for each single-tone modulation (*Figure 5-5*). By using the chart of *Table 5-1*, the amplitude and number of significant sidebands can be found. Since the modulation index equals the frequency deviation divided by the frequency of the modulation, by simply calculating this index, the number and amplitudes of all significant sidebands, as well as the carrier amplitude, can be easily found. Using the calculated modulation index, look under the table's *Modulation Index* column and then across. The relative amplitudes of the carrier and each sideband, as well as their number, are displayed. To find the *bandwidth* of an FM signal, multiply by 2 the number of significant sidebands by the maximum modulating frequency. (*BW = 2N • fm max.*)

Figure 5-7. *An FM signal with a modulation index of 1.5.*

Figure 5-8. *A basic FM superheterodyne receiver.*

Figure 5-9. *Nearly an entire FM receiver on a single chip.*

CARRIER AND SIDEBAND AMPLITUDES AT DIFFERENT MODULATION INDEXES

MODULATION INDEX	CARRIER	PAIRS OF SIDEBANDS TO THE LAST SIGNIFICANT SIDEBAND							
		1ST	2ND	3RD	4TH	5TH	6TH	7TH	8TH
0.00	1.00	--	--	--	--	--	--	--	--
0.25	0.98	0.12	--	--	--	--	--	--	--
0.5	0.94	0.24	0.03	--	--	--	--	--	--
1.0	0.77	0.44	0.11	0.02	--	--	--	--	--
1.5	0.51	0.56	0.23	0.06	0.01	--	--	--	--
2.0	0.22	0.58	0.35	0.13	0.03	--	--	--	--
2.5	-0.05	0.50	0.45	0.22	0.07	0.02	--	--	--
3.0	-0.26	0.34	0.49	0.31	0.13	0.04	0.01	--	--
4.0	-0.40	-0.07	0.36	0.43	0.28	0.13	0.05	0.02	--
5.0	-0.18	-0.33	0.05	0.36	0.39	0.26	0.13	0.05	0.02

Table 5-1. *Carrier and sideband amplitudes at different modulation indexes.*

A *modulation index* of 0 is an unmodulated CW carrier, and produces no sidebands. (*Figure 5-6.*) As the modulation index increases, the sidebands begin to take up more bandwidth, as shown in *Figure 5-7* for a modulation index of 1.5. This is why the frequency of modulation and its amplitude must be controlled in FM communications to lower bandwidth requirements and adjacent channel interference. The two-way (narrowband FM) radio modulation index is typically kept to a maximum of 2 (or 5/2.5) or under, with a maximum allowed frequency deviation of around 5 kHz and a maximum audio modulation frequency of approximately 2.5 kHz — a bandwidth of between 12 and 20 kHz is normally considered adequate to pass sufficient sideband power for full voice intelligibility. In FM broadcast, or wideband communications, maximum deviation is 75 kHz, the maximum audio frequency is 15 kHz, with a deviation ratio of 5 (or 75/15). Maximum BW is 200 kHz, with two 25 kHz guard bands on either side to minimize adjacent channel interference.

5-1C. *FM and AM Comparisons*

There are many advantages that FM has over AM. The most important are much better noise immunity due to limiting employed to remove any AM noise on the received FM signal; and the use of pre-emphasis (boosting the higher frequency at the transmitter) and de-emphasis (attenuating these overemphasized frequencies at the receiver) to lower high frequency noise components. Furthermore, because of the *capture effect*, most interfering signals near or at the same frequency as the desired signal are rejected. Because there is no delicate modulation envelope as in AM, FM does not need linear amplifiers, except in the audio stages, but can use class C high efficiency amplifiers. Also, FM can be modulated by low level modulation techniques, needing very little modulation power — this makes for much better transmitter efficiency. Furthermore, in FM broadcast, no image frequency interference is possible by virtue of its standard high 10.7 MHz IF, so all receivers are typically single conversion types. However, FM does have its disadvantages. Wider bandwidth is needed due to increased sideband production

Figure 5-10. *An FM phase-locked loop demodulator.*

Figure 5-11. *A pulse-averaging discriminator for FM demodulation.*

Figure 5-12. *An FM quadrature detector.*

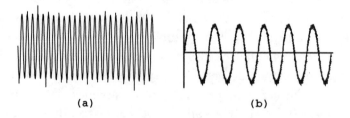

Figure 5-13. *Noise spikes (a) and noise (b) riding on FM waveforms.*

over that of AM. More elaborate and expensive transmitters and receivers are required, mainly on account of the typically higher frequency operation and increased stability requirements over that of AM. (Use is ordinarily restricted to frequencies above 30 MHz because of FM's wide bandwidth requirements.) Also, unless operated through repeaters, FM is a short range, line-of-sight communications medium, since skywave propagation (bouncing the signal off of the ionosphere) causes distortion of the FM wave.

Some important terms to remember:

Center frequency: The frequency of the RF carrier with zero percent modulation. Also called the *rest frequency*.

Frequency deviation: The amount the carrier deviates from its center frequency in *one direction* during frequency modulation.

Frequency swing: The total deviation of the modulated carrier on both sides of the center frequency during frequency modulation.

Modulation index: The ratio between the carrier's instantaneous frequency deviation divided by the instantaneous frequency of the modulation. Used when a single tone at a constant deviation is transmitted.

Deviation ratio: The ratio between the maximum frequency deviation (100% modulation), divided by the maximum audio modulation frequency.

Capture effect: An FM receiver seizes and amplifies the stronger FM signal and rejects another weaker signal at the same or nearby frequencies.

5-2. FM Receivers

Modern FM receivers are usually based on the superheterodyne principle (*Figure 5-8*), and contrast little from AM receivers — except in their method of demodulating and limiting (removing any AM variations of) the incoming FM signal. Furthermore, the use of de-emphasis at the receiver, and the initial pre-emphasis at the transmitter, is also typically found in some FM systems and not encountered in any other form of communications equipment. It is employed to increase the higher frequency but low-in-power audio signals. This allows the audio signals to overcome any interfering noise voltages.

Entire FM receivers are now commonly available on a single chip (*Figure 5-9*), lacking only certain support components and an antenna, an audio power amplification stage, and a transducer (speaker or other device).

5-2A. Detectors

Unlike AM receivers, which detect and amplify the change in amplitude of an incoming signal, a technique of detecting a change in frequency and converting this into a change in amplitude, for reproduction by the speaker or headphones, must be used in FM receivers. The older methods used for this purpose, such as the Foster-Seeley discriminator and the slope, double-tuned, and ratio detectors, are rarely utilized today — more modern techniques have replaced them.

Figure 5-14. *Effect of a high-Q amplifier on AM components.*

Figure 5-15. *A basic fixed-frequency FM transmitter.*

Figure 5-16. *A Pierce varactor (oscillator) modulator for FM production.*

Figure 5-17. *A reactance modulator for direct FM production.*

With a varying input IF, the demodulator (or detector) will output a varying voltage, at an audio rate. However, when the IF is unmodulated, and thus at its center, or rest, frequency, no varying output will occur, consequently creating no amplitude output into the speaker.

Most FM detectors are now usually in IC form. The most common IC detector designs used today are the *phase-locked loop demodulator*, the *pulse-averaging discriminator*, and the *quadrature detector*.

The *phase-locked loop demodulator* (*Figure 5-10*) has become much more common due to the availability of low cost PLL integrated circuits. The PLL receives the FM intermediate frequency into its phase detector, which then compares this incoming signal with the output of the VCO. Since the IF is constantly varying in frequency, if frequency modulation is being received, an "error" voltage, equivalent to this frequency deviation, is outputted from the detector's low-pass filter and DC amplifier as a demodulated audio signal. The higher the frequency deviation, the higher the error voltage outputted, and the louder the subsequent output audio from the receiver's speaker.

A phase-locked loop demodulator needs very little aligning, no exterior inductors and few bias components, and has very low distortion and a great S/N ratio.

The *pulse-averaging discriminator* is a top-of-the-line IC demodulator (*Figure 5-11*). The frequency modulated IF enters the first stage of the discriminator, which is an overdriven amplifier used as a limiter, outputting rectangular and constant amplitude pulses that are equal in frequency to the incoming FM IF. These pulses are then fed into a *monostable multivibrator* (a type of non-sinusoidal oscillator that outputs a single preset duration and amplitude output pulse for each pulse at its input), which outputs its pulse into a low-pass filter. The higher the FM IF frequency (or, to put it another way, the higher in amplitude the original modulating signal), the more pulses outputted per second, and the higher in voltage C_1 charges up to. Because the load is in parallel with C_1, the load will also have this higher voltage across its terminals. The lower the FM IF frequency (the lower in amplitude the original modulation), then the more time C_1 has to discharge across the load, and the lower the voltage will be across that load.

A *quadrature detector* (*Figure 5-12*), while one of the most difficult FM detectors to understand, is one of the most common. The first stage of a quad detector is usually a limiter (which removes AM from the FM signal, lowering noise output), followed by a phase detector (some quad detector ICs do not have a separate limiter, but use a phase detector that has been overdriven, thus easily driving it into saturation and cutoff). The limited FM IF is fed into the input of this phase detector, which acts as an "AND" gate: when both inputs are high, there is an audio output, if one or more is low, then there is no audio output. In other words, both inputs must be of the same phase (same frequency) for any audio output to occur. If there is no modulation present in the IF, then both inputs will be 90° out-of-phase with each other; thus there will be zero output. The phase shift between the inputs occurs due to the reactance of the tuned circuit, consisting of the tunable quad coil and C_2, and its phase relationship with C_1. The tuned circuit is of a very high resistance with a non-reactive nature at resonance (which is at the FM rest, or unmodulated, frequency), and the low value C_1 is, being almost purely capacitive and of a very high reactance to the IF, at a 90° phase shift when compared to input "1", which has a 0° phase shift. As frequency modulation

Figure 5-18. *An automatic frequency control (AFC) used in some non-crystal FM transmitters.*

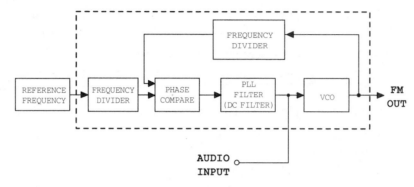

Figure 5-19. *An FM phase-locked loop modulator.*

Figure 5-20. *A phase modulated (indirect FM) transmitter.*

Figure 5-21. *Principle of phase modulation.*

Figure 5-22. *A basic transistor phase modulator.*

Figure 5-23. *A practical phase modulator using a FET.*

occurs in the IF, the phase relationships between the inputs change. This is because the tuned circuit goes off resonance due to this change in frequency, which alters the reactance of the tank, changing the phase relationship between the tank and C1 and, consequently, input 2 as compared with input 1. The output is a series of pulses into a low-pass filter, which vary in width depending on the frequency swing of the frequency modulated IF. The pulses are averaged and the resultant audio is output across the next stage. Tuning for rest frequency is accomplished by tweaking the external L_1.

5-2B. *Limiters*

A separate limiter can be used in front of any detector that does not have built-in limiting. The limiter is utilized to remove most of the noise-generating AM components (*Figure 5-13*), or *incidental AM*. A limiter is basically an overdriven class A single ended amplifier with a tuned output circuit, used after the last IF stage of an FM receiver, to clip off any amplitude variations in the received signal. The limiter cuts-off the amplitude peaks of the signal, in some cases creating a ragged square wave. The signal is turned back into a sine wave by a tank circuit, which filters out the harmonics caused by the clipping and recreates the original sine wave.

A simple tank circuit, itself in class C amplifiers, will also act as a limiter by removing any RF or IF amplitude variations — if the Q of the tank is high enough. A high Q tank will be narrow enough to only pass the desired frequency variations and reject any amplitude variations, which are riding on the FM waveform, as being outside of its bandwidth, thus removing most of the incidental AM (*Figure 5-14*). Also, another way to think of the effect a high Q tank has on amplitude variations is that the tank will not lose enough energy in the flywheeling tank in time for the next pulse, which may possibly vary in amplitude. This amplitude variation will not be noticed because the high Q tank continues to "ring", so any lowering of signal amplitude will not be outputted to the next stage.

5-3. FM Transmitters

FM transmitters (*Figure 5-15*) can be significantly different in design than the average AM transmitter — low level modulation is always commonly used with class C frequency multipliers, employed even after modulation; and most FM transmitters are designed to function only in the VHF and higher area of the radio spectrum.

5-3A. *Modulation Circuits*

An FM transmitter must use some technique to vary the output frequency in response to amplitude variations in the baseband signal, while allowing little or no amplitude variations in the carrier. There are two basic ways to perform this function: *direct*, or frequency modulation, which directly changes the frequency of the carrier oscillator in proportion to the amplitude of the baseband signal; and *indirect*, or phase modulation, which indirectly creates FM by varying the phase (and thus the frequency) of the carrier oscillator's output frequency.

Direct modulation can be created by the use of a *varactor modulator*. Varactor modulators use varactor diodes to vary the frequency of an oscillator to create FM. Varactors, also referred to as varicaps, work on the same principle as the junction diode. When a reverse bias is placed across a semiconductor diode, the depletion region widens. Think of this region as the dielectric, since it acts as an insulator. The thicker the dielectric, the less capacitance. So, increasing reverse bias on any semiconductor diode decreases its capacitance, and decreasing the reverse bias decreases the depletion region (equivalent to the plates being placed closer together) increases capacitance. A varicap is a diode that is optimized to perform this very important function.

A common FM modulator is the *Pierce varactor oscillator* as shown in *Figure 5-16*; but the same idea can be applied to any *Hartley-* or *Colpitts*-based oscillator. Resistors R1 and R2 set the Pierce varactor's reverse bias, and thus the oscillator's center, or rest, frequency. When an audio signal is applied at the oscillator's input port, it adds to or subtracts from the varactor's reverse bias, which changes the capacitance of the circuit, altering the modulator's center frequency (FM). A varicap that is in series or parallel with an oscillator's crystal can only deviate the center frequency by a few hundred hertz, and is therefore used mainly in narrowband FM applications. Still, to increase both the frequency and the value of deviation, one or more stages of frequency multiplication may be utilized.

Reactance modulators (*Figure 5-17*) can also be employed to generate direct FM. They are used to vary the frequency of the master oscillator, but cannot typically be used in *crystal controlled* master oscillators, or MOs. These modulators are placed across the LC frequency determining tank of the master oscillator. An audio signal is then fed into the reactance modulator, which is biased so that it appears to be *either* a purely capacitive or, in certain circuits, as a purely inductive reactance to the MO's tank. This audio signal varies the current through either a transistor, FET, or tube. As the current of the active element increases and decreases, the reactance of the active element changes, varying the MO's tuned circuit reactance in equal proportion, thus altering the output frequency of the MO around its resting frequency, creating direct FM.

The reactance modulator is capable of very wide frequency deviations when used in LC oscillators, as well as having inherently very little distortion. However, since it cannot control a crystal MO, it must use other frequency stabilizing circuits, such as AFC (*automatic frequency control, Figure 5-18*). FM transmitter AFC inspects the output from the frequency multipliers (or the output directly from the MO), and sends this sample into the mixer, comparing it with a crystal oscillator reference frequency. If the carrier frequency is found to be off of its design frequency, the mixer will output the difference between the two frequencies into the discriminator. The discriminator will then output a correction voltage (since this new frequency will be different from its preset rest frequency) into the reactance modulator, changing its reactance, which forces the MO back on frequency. Additionally, a reactance modulator must be designed to be *purely* reactive, or it will introduce AM components that, when transmitted, cause noise at the receiving end.

A *phase-locked loop* can also be used to generate direct FM (*Figure 5-19*), and is employed in many of the newer frequency-synthesized transceivers. The PLL produces and maintains a rock steady rest frequency, while any modulation injected at the input to the voltage controlled oscillator (VCO) varies the PLL modulator's output frequency in step with the modulation voltage, thus creating FM.

Phase modulation (PM), or indirect FM, can be used to produce frequency modulation. In many FM applications it is employed more than direct FM because of PM's ability to utilize a crystal oscillator or frequency synthesizer as the MO, thus giving the FM transmitter the ability to maintain a very accurate and steady output frequency — especially in narrowband use. PM has limited frequency deviation abilities, however, when compared to a reactance modulator, so it is ordinarily used with large amounts of frequency multiplication. The *output* of the transmitter's MO (*Figure 5-20*), or its buffer, is modulated, and not the input as in direct FM. This allows the MO to be crystal or PLL controlled, as well as protecting the MO from any deleterious effects of the modulation — resulting in an extremely stable oscillator. The phase of the RF carrier from the MO or its buffer is changed, creating a leading and lagging time difference between the resting carrier frequency and the modulated carrier frequency; which is the same as varying the frequency of the carrier.

In *Figures 5-21* and *5-22* two simple PM circuits are shown. Recall that an RC network can alter the output phase angle of any input signal, inasmuch as a pure resistance has a phase angle of $0°$, while a pure capacitance has a phase angle of $-90°$ — and combining the two in a circuit will create a phase angle anywhere between $0°$ and $-90°$, depending on the values of the resistance and the capacitive reactance. In *Figure 5-21* the output will lead the input by any value between $0°$ and $-90°$ by varying the ratio between X_C and R. If, by the application of a modulating voltage, either the value of the resistance or the capacitive reactance could be changed, then phase modulation would take place. This can be accomplished by replacing the resistor with a FET or a bipolar transistor (*Figure 5-22*), which acts as a voltage or current controlled variable resistance. As the modulating signal amplitude increases and decreases at the input, the transistor's resistance decreases and increases respectively, varying the ratio between the resistive element (the transistor) and the reactive element (the capacitor), thus changing the phase angle of the output. Since the output is also amplitude modulated due to the varying audio bias on the gate of the active component (the transistor), the stages after the modulator are biased class C, which removes any generated AM components. (See *Limiters*.)

A FET-based phase modulator is shown in *Figure 5-23*, with the audio input injected into the low-pass filter consisting of R_1 and C_2. This filter is used to remove PM's inclination to increase the carrier's frequency deviation as the frequency of the modulating signal increases. The MO rest frequency is placed at the gate of the FET, and through C_3 and across the tank, which consists of L and C_5. Through the interaction of C_3, the tank and the FET, PM (as well as AM) is created. The amplitude modulation is removed by class C amplifiers or frequency multipliers placed after the modulator.

A varactor diode and a resonant tank can also be used to modify the phase of the output of the MO (*Figure 5-24*). When the audio amplitude across a varactor diode changes, the capacitance of the varactor, and thus the resonant frequency of the tank, is varied, creating phase modulation. The tank consists of L_1, C_3, and the varactor, D_1. The varactor's bias is then added-to or subtracted-from by the modulating audio signal while the carrier is sent through C_4 and into the tank. The tank, due to this changing bias on the varactor, alters its capacitance, and thus the resonant frequency of the tank circuit. When the tank is on frequency, then the circuit is purely resistive with no reactive components, since both X_C and X_L cancel each other at resonance — with a resultant phase angle of zero. When the audio signal across the diode is changed the entire reactance of the tank is altered, and will become either

Figure 5-24. *A varactor phase modulator.*

Figure 5-25. *A class C frequency multiplier.*

Figure 5-26. *An FET push-pull frequency doubler.*

capacitive (if the frequency decreases), or inductive (if the frequency increases), producing lagging and leading phase shifts. The bias dropped across R2 sets the diode's quiescent (rest) reverse bias, which fixes the tank's resonant frequency to that of the unmodulated carrier. R_1 and C_1 form a low-pass network, which removes PM's proclivity to increase the output frequency swing for the higher audio modulation frequencies, while the RFC is used to minimize the loading of the tank, which would decrease its Q. This phase shifter is also followed by either class C amplifiers or frequency multipliers to remove any amplitude variations produced by the phase shifting.

5-3B. *Frequency Multipliers*

Frequency doublers and triplers, used extensively in FM transmitters, are typically basic tuned output nonlinear (class C) amplifiers. They are utilized to multiply the frequency modulated carrier obtained from a stable crystal oscillator (the master oscillator) that would have difficulty working dependably at such high frequencies. As well, multipliers are able to increase the deviation of any frequency modulation present. A carrier oscillator's frequency and its deviation might be multiplied by thirty or more times, as needed. A carrier frequency that started out at 6 MHz with an FM deviation of 150 Hz would be, if multiplied by a final 30X chain of multipliers, an output frequency of 180 MHz with a deviation of 4,500 Hz.

Figure 5-25 shows a common class C amplifier functioning as a frequency multiplier. Distortion of the fundamental is created by the nonlinear operation of the class C amplifier, thus generating a signal that can be rich in harmonics. The input tank is tuned to the fundamental, while the output tank is tuned to an exact harmonic of this frequency. The output tank not only passes the chosen harmonic on to the following stage, but also attenuates the now unnecessary fundamental. Another type of frequency multiplier is the *push-push doubler* (*Figure 5-26*). One FET receives a 180° out-of-phase signal input as compared to the other FET — yet their outputs are tied together (in parallel), so twice as many pulses occur into the output tank circuit as would occur in a normal push-pull circuit. Because the tank is tuned to the second harmonic, a frequency multiplication then takes place. Another common method of frequency multiplication is the *tripler varactor frequency multiplier* (*Figure 5-27*). As the input signal is coupled into the input tank, which is tuned to the fundamental frequency, the varactor is successively switched on and off, which distorts the input signal, creating harmonics. The second harmonic is sent to ground through the series bandtrap filter, while the third harmonic is passed on to the secondary of the 3X fundamental tuned tank circuit.

Frequency multiplier amplifiers, despite their high frequencies, do not typically need *neutralization*, since their input and output circuits are tuned to different frequencies.

5-4. Selective Calling

Selective calling is a method that can be employed in FM to keep all receivers in a network "off" until it is desired to contact all or some of these receivers. This function can be performed by the use of CTCSS, or *continuous-tone coded squelch system*. Many FM transceivers use *channel guard* and *private line*, which are proprietary selective calling systems used to "unsquelch" the receiver's audio

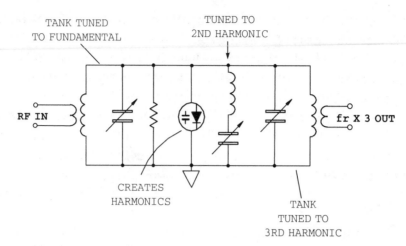

TANK TUNED
TO FUNDAMENTAL

TUNED TO
2ND HARMONIC

RF IN

fr X 3 OUT

CREATES
HARMONICS

TANK
TUNED TO
3RD HARMONIC

Figure 5-27. A tripler varactor frequency multiplier.

Figure 5-28. CTCSS tone generation in a transmitter.

Figure 5-29. Detection of the CTCSS tone in a receiver.

Figure 5-30. A pre-emphasis circuit in an FM transmitter.

Figure 5-31. A de-emphasis circuit in an FM receiver.

section by the caller's transceiver transmitting a continuous sub-audible tone on the RF carrier below 300 Hz, which is under the passband of the receiver's audio amplifiers, or tone-squelch burst codes of anywhere between 1.6 to 2.55 kHz. *Figures 5-28* and *5-29* show the general idea of CTCSS. At the transmitter, the sub-audible tone at a very low frequency (a few hertz, depending on frequency multiplication) is injected, along with the audio modulation, into the modulator, where it is sent to the frequency multipliers and then to the finals and out to the antenna, where it is radiated. The receiver picks up the original audio as well as the sub-audible tone. The audio is sent on to the audio amplifiers and speaker, while the sub-audible CTCSS tone is tapped off and sent to a bandpass filter to separate its frequency from any other tones present, then to a rectifier and low-pass filter, where it is changed into a DC control voltage to switch off the squelch, which activates the audio amplifiers. The operator is now able to listen to the open channel.

As mentioned above, another type of selective calling uses bursts, a pulsed tone, or a combination of tones, usually sent only at the beginning of a transmission, at frequencies anywhere between 600 to 3,000 Hz (depending on the system), to open the squelch and allow reception of the FM signal.

A few things must be kept in mind when working on a CTCSS equipped transceiver. When aligning or testing an FM transmitter section, the sub-audible tone will be modulating the carrier, whether or not there is any speech modulation present, confusing any desired readings if it is not expected. Some CTCSS produce an FM deviation of approximately 750 Hz, thus allowing a maximum narrowband deviation of the speech component of only ±4.45 kHz. Also, when testing a CTCSS receiver, a CTCSS tone generator (or *CTCSS encoder*) must be used to open the squelch (unless there is a switch to turn CTCSS on and off) or the receiver might appear dead.

5-5. Pre-Emphasis and De-Emphasis

Pre-emphasis and *de-emphasis* is a technique employed in some FM systems to lower noise as heard in the receiver. Pre-emphasis (*Figure 5-30*) overamplifies the higher audio frequencies in a stage just before modulation in the transmitter. Generally a simple high-pass filter is used. Since higher frequency components have less energy in speech and music than lower frequency components, the higher frequency will now not be masked by either internally or externally generated noise. At the receiver, de-emphasis (*Figure 5-31*) is performed on the audio, right after the demodulator, which attenuates the high frequencies created by the pre-emphasis back to natural levels. This also, of course, significantly lowers the amplitude of any high frequency noise present in the signal.

5-6. FM Transceiver

An entire no-frills FM hand-held transceiver is shown in *Figure 5-32*. Most of the circuits of this chapter, as well as many of the support circuits explained in *Chapter 11*, are in this block diagram. For transmission, the *PTT (push-to-talk)* button is depressed, supplying power to the transmitter section and removing power from the receiver. As the operator speaks into the microphone, the transmitter's 1st audio voltage amplifier increases the very low audio voltage, which is then further boosted by the

Figure 5-32. A complete FM hand-held transceiver.

second audio amplifier while also furnishing amplitude limiting. The low-pass filter (LPF) stage supplies frequency limiting and removes any harmonics created by the prior amplitude limiting. Maximum frequency deviation and significant sideband generation and spacing, and thus bandwidth, are consequently restricted. This limited audio is fed into the varactor modulator, where the bias on the varactor, and thus the frequency of its oscillator, is varied by the audio voltage. The oscillator's output tank can be tuned to some multiple of the varactor's crystal oscillator (See *Pierce varactor oscillator*) to increase its output frequency (frequency multiplication). Its output frequency can be varied by the channel select switch, which allows the operator to switch in different frequency-determining crystals — or a PLL can be used. This frequency is outputted into a frequency tripler amplifier to increase the carrier frequency, as well as any deviation, and then sent into a buffer amplifier. The buffer amplifies, filters, and isolates the varactor modulator's oscillator and tripler amplifier from the next stage, the transmit mixer. This frequency, as well as the offset oscillator's frequency, is mixed in the transmit mixer, creating sum and difference frequencies. The offset oscillator is able to choose from three crystals, switched in by the operator, which allows the transmitter to be either at the same frequency as the received signal (referred to as *simplex operation*, or when communication is directly from transceiver to transceiver), or to offset the transmit frequency from the receive frequency by ±600 kHz for *repeater* utilization. (A repeater receives on one frequency and retransmits your signal on another.) The sum of the mixing products is filtered by the BPF (bandpass filter) and fed into the predriver, the driver, and the power amplifier. The output of the power amplifier is sent into the diode RX/TX selector, which consist of diodes that are forward biased by a DC voltage, allowing the transmitter's signal to pass, while other diodes are reversed biased into the receiver section, disconnecting the receiver on transmit. The LPF and *wavetrap* further attenuate all harmonics, as well as assure that the most powerful harmonic, the second, has added attenuation. When the PTT button is released, the receiver section is supplied with power, while power to the transmitter section is removed. The received signal passes through the LPF and wavetrap filter, which protects the receiver from overload due to strong adjacent UHF signals. The receiver's power supply forward biases other diodes in the diode RX/TX selector, allowing the received signal to pass through to the first RF amplifier, while blocking any signal from entering the transmit section. The high frequency first and second RF amplifiers provide a limited amount of gain and selectivity. The amplified RF signal is sent into the receiver mixer, where it mixes with the output of the combined local oscillator/varactor modulator. The difference frequency is filtered and narrowed by the IF crystal filters and sent into the IF amplifiers, then into the combination IF amp/detector/limiter, where the IF is further amplified as well as limited to remove AM noise, and demodulated. There are two outputs from the amp/detector/limiter — one to the audio amplifiers and one to the squelch section. The squelch section compares the incoming signal level with a preset reference voltage to discern if the signal is of an adequate amplitude to turn on the receiver's audio amplifiers. If the signal is of sufficient amplitude the audio amplifiers are turned on — and the detected audio is amplified and fed to the speaker for reception by the listener. An additional circuit, the *battery saver* stage, rations battery power during periods when no or only low level signals are being received. It pulses the receiver on and off five to ten times a second until the squelch circuit detects that a sufficient signal has been received, at which time the power becomes continuous.

CHAPTER 6
Single-Sideband Suppressed Carrier

As lower cost transmission and receiving equipment for the generation and reception of single-sideband suppressed-carrier signals have become available, it has turned into the preferred method for long range communications in specialized and amateur applications. It is also used for voice signals all the way from low frequency up to UHF.

6-1. SSB Fundamentals

Single-sideband suppressed-carrier is a type of amplitude modulation that transmits only a single sideband, while the carrier and the other sideband are heavily attenuated (*Figure 6-1*). This is a huge improvement over standard AM, in which the carrier (which contains no information and is thus useless) and the two sidebands (each containing duplicate information) are transmitted. SSB also needs only half the frequency spectrum as standard AM for transmission, since one sideband is suppressed. The power conserved is not transmitting the carrier and one sideband can be better utilized in transmitting the single-sideband signal. This also means less noise is received due to this decrease in bandwidth. (*Figure 6-2.*) Fading is less of a problem because complex phase relationships between all of the transmitted elements (both sidebands and the carrier) need not be maintained, considering only one sideband is being transmitted. As well, a transmitted signal is only generated when modulation is present, as the carrier is suppressed within the transmitter, saving even more power.

6-1A. SSB Modulation

Single-sideband is produced at the SSB transmitter (*Figure 6-3*) by injecting a carrier and a modulating audio signal into a *balanced modulator*. The balanced modulator mixes these inputs, which results in lower and upper sidebands, while suppressing the carrier through phase cancellation or common-mode rejection techniques. The balanced modulator then outputs the resultant double-sideband suppressed carrier (DSB) signal to the next stage. Keep in mind that the original carrier must mix with a modulating signal to produce any sidebands. Without modulation present there will be no output — a sideband must be created for SSB transmission, since the carrier itself is severely attenuated within the balanced modulator. The next stage can consist of highly selective filters to pass either the upper or lower sideband or, through phase-cancellation methods, attenuate the undesired sideband using twin balanced modulators and phase-shifter circuits. As only low-level modulation methods are put to use in SSB, linear amplifiers must be utilized throughout to maintain the delicate modulation envelope. At the SSB receiver (*Figure 6-4*), the signal is amplified and inserted into a nonlinear mixer, called a *product detector*, where it is

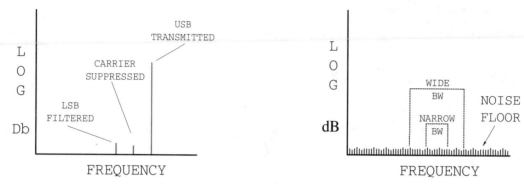

Figure 6-1. The frequency spectrum of an SSB signal.

Figure 6-2. A narrower SSB bandwidth over standard AM equals less received noise.

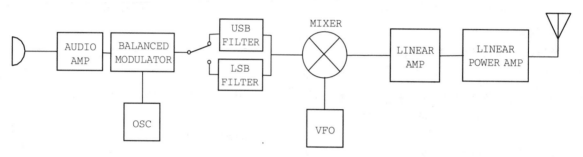

Figure 6-3. A basic filter-type SSB transmitter.

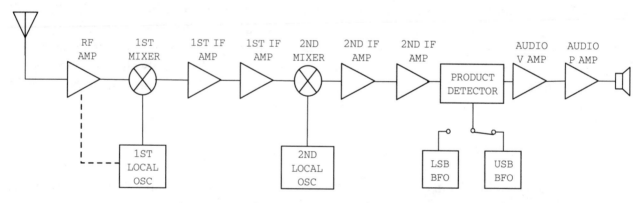

Figure 6-4. A dual-conversion SSB receiver.

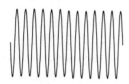

Figure 6-5. A single-tone SSB signal.

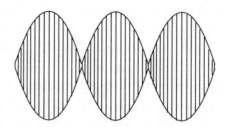

Figure 6-6. A two-tone SSB signal showing the modulation envelope.

mixed with the output of a *carrier oscillator* (or *BFO*) to supply the missing carrier. The output is an audio signal, which is then amplified and sent to the speakers.

When single tone modulation is applied in SSB, a constant amplitude and frequency (CW) signal is produced, as in *Figure 6-5*. A two-tone baseband signal produces the classic SSB modulation envelope oscilloscope display of *Figure 6-6*, with the amplitude of the RF modulation envelope contingent on the audio modulation level. When overmodulation occurs, the two-tone signal begins to flat-top (*Figure 6-7*), creating severe distortion and spurious outputs.

6-1B. SSB Output Power

In SSB, as in AM, the peak envelope power (PEP) is the measurement of the peak (maximum) instantaneous power, with 100% modulation applied, of the transmitted signal, and is calculated by V^2_{RMS}/R, or $V_{RMS} \times I_{RMS}$, or $I^2_{RMS} \times R$, of the peak values at 100% modulation.

Unlike AM, there can be no power measurements without modulation, since there is no output possible with SSB unless the transmitter is modulated and the sidebands are produced.

6-1C. Standard AM and SSB Comparisons

SSB transceivers have become the preferred method of communicating at extreme distances, and have now virtually replaced all AM transmitters for long-range two-way voice communication. SSB generation and reception, however, demands more expensive and complex equipment than AM. The SSB transmitter is low level amplitude modulated, but circuits must be included that suppress the carrier and filters out the extra sideband, while the receivers must contain added circuitry that reinjects the carrier, at a very steady frequency, before the baseband signal is demodulated.

6-2. SSB Receivers

An SSB receiver (see *Figure 6-4*) is basically the same as any superhet receiver. The main difference is in the detection of the SSB signal as well as a few other design features: Some SSB receivers use a *wideband* first IF with a fixed frequency crystal, LO, that passes many frequencies, and tune for the desired signal through a VFO (*variable frequency oscillator*) during the second IF conversion (See *Dual-Conversion*). Older receiver units also use some type of *varactor* controlled AFC to control drift in either the first or second LO, whichever section uses a VFO. The SSB receiver of *Figure 6-4* is a double, or dual, conversion receiver. The low level RF signal is picked-up by the antenna, boosted by the RF amplifier, and injected into the 1st mixer where it is heterodyned with the first LO to form the *high image-frequency*, rejecting first intermediate frequency. This is outputted, filtered, and sent into the IF amplifiers to increase gain and improve selectivity. This first mixer stage typically has a LO that uses a VFO or frequency synthesizer, since it must normally be tunable. The first IF is then injected into the second mixer, which converts and filters it into a lower second IF. The second IF amplifiers supply most of the receiver's selectivity and sensitivity. The second LO is normally a fixed frequency crystal

Figure 6-7. *Overmodulation of an SSB two-tone signal.*

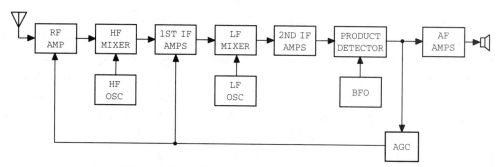

Figure 6-8. *AGC in a SSB receiver.*

Figure 6-9. *A diode-ring balanced modulator for SSB.*

Figure 6-10. *The differential amplifier balanced modulator.*

Figure 6-11. *A type of crystal-lattice sideband filter.*

oscillator. This second intermediate frequency is then sent into one input port of a balanced modulator (BM). The balanced modulator is used to detect the SSB signal after reinsertion of the carrier by either the LSB or USB BFO (*beat frequency oscillator*), depending on which sideband is the desired one for reception, which is injected into the second BM input port. The BM's output port sends the resulting difference frequency, which is now an audio signal, into the audio amplifiers and on to the speaker.

6-2A. SSB Detection

When a balanced modulator is utilized for this purpose of SSB detection, it is frequently referred to as a *product detector*. Since a balanced modulator is nothing more than a nonlinear mixer, many other types of nonlinear mixers can be employed for this purpose.

Considering the carrier is missing from the received signal, a new carrier must be inserted at the same frequency as the missing carrier would have been after the second conversion in a dual-conversion receiver, or after the first and only conversion of a single conversion receiver. This new inserted carrier frequency will be at a much lower value than the one that would have been transmitted, since it would have gone through one or two down-frequency conversions. An oscillator called a BFO generates this missing carrier. The IF, with the audio information being carried on its sidebands, is injected at one input of the product detector, while the BFO frequency is injected into the other. The output of the product detector is the low difference frequency of the beating of these two signals, which is a reproduction of the original audio. The other non-audio signals are eliminated by different means, depending on the type of product detector used. The BFO frequency can be suppressed by the canceling action of the "balancing-out" of the inphase reinserted carrier when a balanced modulator is employed, or through low-pass filtering if standard mixers are utilized — while the IF and the sum frequency can be passed to ground by the LPF. Any frequency higher than the desired audio is out of the passband of the audio amplifiers, further helping to attenuate any high frequencies components. The audio is sent to the audio amplifiers and on to the speaker or headphones. The BFO itself is normally a crystal oscillator, since extreme stability is vital: If the BFO is off frequency by more than 100 Hz, or its output amplitude is not substantially larger than the received sideband, the resulting signal becomes difficult to read. Most receivers use twin switchable BFOs so that the desired sideband, either upper or lower, can be chosen for amplification to the speaker. Furthermore, a small variable tuning capacitor or varactor circuit can be utilized to slightly tune the BFO up or down to vary the tone of the received audio for increased clarity. This is called RIT (*receiver incremental tuning*) or *clarifier control*.

6-2B. SSB AGC

Most AGC in an SSB receiver is similar to other RF receivers. The detected audio is tapped after the product detector (audio-derived AGC), filtered to DC and amplified, and used to control the gain of the RF and IF amplifiers (*Figure 6-8*). Most AGC in SSB is switchable for a long or short attack time, depending on whether speech (long pauses between words) or Morse code (short pauses between "words") is being received. This is normally accomplished by using different RC time constants and switching in different value resistors. A low value resistor is adopted for CW reception to remove the AGC control voltage quickly; and a high value resistor for SSB reception to hold its AGC control voltage for a longer

Figure 6-12. *Another common crystal-lattice sideband filter.*

Figure 6-13. *A crystal ladder sideband filter.*

Figure 6-14. *A mechanical resonant sideband filter.*

time period. Also, *delayed AGC* is common in SSB. This type of AGC does not reduce the gain of any stage unless a sufficiently strong signal is received, avoiding the problem of lowered gain even with weak input signals. A Zener can be employed in this role, and will hold the AGC voltage to some constant value until a signal of adequate strength is received to activate the AGC circuits. Other SSB receivers utilize an AGC that is tapped off of the second IF, amplified, rectified, filtered to DC, further amplified, and applied to the appropriate RF and IF stages.

6-3. SSB Transmitters

The simple SSB transmitter of *Figure 6-3* utilizes a crystal oscillator to supply the stable low frequency carrier. SSB is almost always generated at a much lower frequency than the transmitted RF carrier frequency that is finally injected into the antenna. This assures easier filtering of the unwanted sideband, while the carrier is increased in frequency by heterodyning in a later stage. The carrier and audio frequencies are first fed into a balanced modulator, which is nothing more than a doubly-balanced mixer, that suppresses the carrier and the audio frequencies, while outputting the resultant upper and lower sidebands created by the nonlinear mixing action. Doubly-balanced mixers offer high port-to-port isolation, which allows for almost no coupling of the separate input frequency's (LO or BFO, and RF and IF, and AF) energy between the input/output ports. The carrier's input into the balanced modulator is usually about ten times the amplitude of the audio signal for distortion reduction. Then, at the output of the balanced modulator, either the LSB or the USB can be chosen, depending upon the operator's preference, by switching in a very sharp filter, which sends the selected sideband to a transmit mixer for frequency translation into a higher frequency. The VFO for the mixer is commonly a PLL or DDS frequency synthesizer, designed to supply a wide tuning range for the preferred output operating frequency. A tuned circuit at the output of the mixer can be set to attenuate either the resulting sum or difference frequency. Ordinarily, another tuned circuit in a tuned linear amplifier stage supplies further attenuation of the unwanted sum or difference frequency, while providing amplification for the desired frequency. These amplifiers are commonly biased class B linear or, more rarely, class A. The sideband signal is then further amplified by the final linear push-pull power amplifier and sent into the antenna, where it is radiated.

6-3A. Balanced Modulators

The most common balanced modulator is the *diode-ring* type of *Figure 6-9*. Between two center-tapped input and output transformers are bridge-connected diodes. The carrier is inputted at the primary of T_1, while the audio is injected at the center tap of T_2. The carrier and audio are suppressed in the modulator, with the double sideband (DSB) output taken from T_2's secondary. The carrier is suppressed by the action of T_2: the diodes force the carrier current to flow equally between the upper and lower sections of the primary of T_2 and out-of or into the grounded center tap, creating equal but opposite magnetic fields in each section of the winding between T_2's grounded center tap. This causes the carrier to be canceled, or balanced out — thus no carrier energy is coupled to the secondary of the transformer. Obviously, for proper carrier suppression, the components of a balanced modulator must be perfectly matched. Balanced modulators are also available in module form, with the diodes and transformers already matched internally; or as an IC using *differential amplifiers* (*Figure 6-10*). The IC differential

Figure 6-15. *A flexure-mode mechanical sideband filter.*

Figure 6-16. *A single-filter SSB transmitter.*

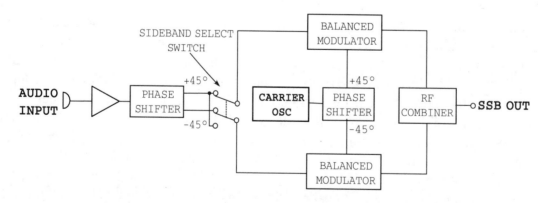

Figure 6-17. *A filterless phasing-method for SSB production.*

amplifier rejects the common mode RF carrier signal because there is no difference in potential between the outputs of transistors Q_1 and Q_2 for the inphase carrier. The modulation injected into the base of the constant-current amplifier Q_3 disturbs the balance of the circuit, creating nonlinear mixing between the carrier and the modulating audio, producing both upper and lower sidebands and suppressing the carrier. (See *Differential Amplifier.*) Carrier suppression with the average balanced modulator is 40 dB, while suppression capabilities of up to 60 dB are not uncommon.

6-3B. *Sideband Attenuation*

The undesired sideband out of the balanced modulator can be attenuated in one of two ways. The desired sideband can be selected and sent on for amplification by the application of a *sideband filter*, while the undesired sideband is suppressed because it is out of this filter's passband. There are two types of crystal filters commonly used for this purpose: the *crystal-lattice* type (*Figure 6-11* and *Figure 6-12*) and the *crystal ladder* type. (*Figure 6-13*. See *Filters: Crystal and Ceramic* for a further explanation of these two configurations.) Also, expensive and low frequency (under 600 kHz), but high in quality (very narrow and steep bandpass characteristics), *mechanical resonant* (*Figure 6-14*) type filters are found in a few new SSB transmitters and many older units. The mechanical filters consist of metal disks and rods in a single sealed assembly, with the DSB signal fed into an input coupling coil, which has been placed in the permanent magnetic field of a biasing magnet. An induced magnetic field produced by the DSB input signal either adds to or opposes the permanent magnetic field, producing movement of the coupling coil and its attached transducer rod. The resonant disks are coupled together by connector rods, with the first and last resonant disks attached to the transducer and connector rods. The resonant disks, which are cut to exact dimensions, will only vibrate at their natural frequency, which is dependent on their diameter and thickness. If the frequency at the filter's input is at the natural resonant frequency of the discs, the resulting vibrations will vibrate the attached output transducer rod, which will move the output coil in synchronicity. This coil is also in the magnetic field of a permanent magnet, so its movement produces an induced output signal voltage that is a duplicate of the input resonant frequency. The sideband that is not within the passband of this filter is severely attenuated. An additional type of mechanical filter is called the *flexure-mode mechanical filter* (*Figure 6-15*). An input signal is injected into a ceramic crystal transducer, which converts this varying signal voltage into a mechanical vibration. The vibrations are conveyed to an iron-nickel alloy bar on which the crystal is attached. This alloy bar twists, transferring this flexing to another crystal and alloy bar combination through coupling rods, creating a vibration in the output piezoelectric crystal, which causes a corresponding output signal voltage at this filter's resonant frequency.

Another technique of producing SSB uses two carrier frequencies and one SSB filter (*single-filter* SSB, *Figure 6-16*), with the carrier frequency shifted to send one sideband through the sideband filter. The other sideband is out of the filter's passband and is consequently suppressed.

An additional manner of suppressing both the carrier and one of the sidebands is the *phasing method* (*Figure 6-17*) for SSB generation. This arrangement uses a quadrature phase shifter to break the audio into two constituents that are out of phase by 90°. The carrier is also split and phase shifted by 90°, with one -45° audio and one -45° carrier constituent inserted into one of the balanced modulators, while the

+45° audio and +45° carrier constituent is inserted into the other balanced modulator. The balanced modulators attenuate the carrier and outputs a DSB signal to the RF combiner, which adds the two signals. Since there is a phasing difference between these input signals, one sideband is intensified, while the other is canceled. The output is SSSC.

CHAPTER 7
Other Communication Applications

There have been various processes adopted that employ the basic AM, SSB, FM or PM communication techniques that enable us to increase the reliable range and/or information transfer rate of an RF signal over that of common simplex single-channel radio communication methods. Additionally, small, low cost, and simple-to-operate equipment has opened up long and short range RF communications to the untrained operator. Many of these methods are relatively new, while some have been available for years. *Multiplexing* is important for conserving spectrum space and permits economical multichannel communications to take place over a single transmission path. *Data communication* techniques are the perfect medium for efficient digital and analog high speed and high quality information transmission. *Satellite* use has revolutionized almost all forms of voice, data, and video long-distance communications. *Spread spectrum* is a developing and expanding form of RF communications that can be reliably employed in a very crowded spectrum with little signal degradation. *Cellular radio* has made instant mobile telephonic wireless communications a low-cost and viable communications medium. *CW* is an old and dependable long-range communications means that is still quite popular. *Repeaters* have made low-powered VHF (and above) terrestrial line-of-site communications possible over long distances, especially for FM.

7-1. Multiplexing

Multiplexing refers to the ability to send two or more channels of information over a single channel. There are two methods of communication multiplexing: *frequency-division multiplexing* and *time-division multiplexing*.

7-1A. *Frequency-Division Multiplexing (FDM)*

FDM has the ability to transmit multiple signals over one wide channel by converting each desired signal into a different narrow frequency span, and transmitting them all simultaneously within that single wide channel (*Figure 7-1*). Bandpass filters are used at the receiving end, set to the proper frequencies, to remove the desired intelligence from the appropriate narrow frequency channels.

In an FDM transmitter (*Figure 7-2*), each baseband voice signal modulates a different carrier. This modulated carrier is then fed into a linear mixer, which does not create sum and difference frequencies, but allows the original signal frequencies to be maintained and separated without generating any new signals. This linear mixing action combines all the channels into a single complex signal, which is then fed into a wideband transmitter and sent out over the air on one broad channel. AM, FM, PM and SSB can be used to modulate the baseband signals.

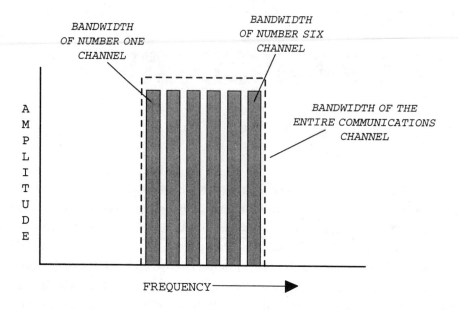

Figure 7-1. *The frequency spectrum of an FDM signal.*

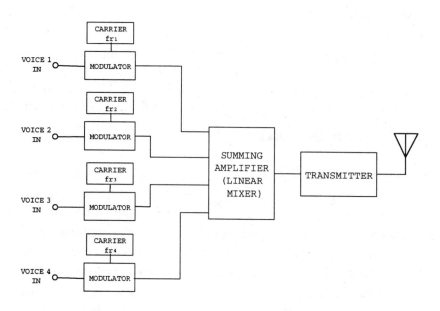

Figure 7-2. *A four channel FDM transmitter system.*

The FDM wideband receiver (*Figure 7-3*) amplifies and demodulates this signal, while the bandpass filters are each tuned to a specific carrier. The filters pass these modulated carriers on to the detectors, which demodulate and output the original baseband signals to the appropriate transducers. FDM is typically used for analog information transfers.

7-1B. *Time-Division Multiplexing (TDM)*

TDM transmits one signal in one time period, and another signal in another time frame, over a single channel. In a TDM transmitter a repetitive time period is reserved for each specific voice or data channel. It is basically a time sharing approach. One channel of information is sent, then the next, then the next, until it returns to the first channel and the process repeats — all on a single radio channel. A very common modulation technique for TDM is the digital method called PCM. (See *Pulse-Coded Modulation.*) To transmit voice (*Figure 7-4*), the analog waveform is sampled at set time periods by an analog-to-digital converter (ADC), which outputs a parallel digital word for each discrete analog amplitude level. This is then sent to a parallel-to-serial data converter and on to a multiplexer, which relays the PCM, at repetitive time slices, through to the transmitter. The multiplexer can be compared to an electronic rotary switch which is programmed by the timing and control circuits to place a digital word, waiting at the input of the multiplexer, to its single output. The timing and control circuits then instruct the multiplexer to send the next word, waiting at the next input, to its single output. This process continues until all of the inputs have been serviced; then the cycle repeats. The transmitter converts this information into AM or FM PCM, and supplies synchronizing pulses to help the receiver properly decode the PCM. The transmitter thereupon sends the signal out over the air in AM constant amplitude pulses (no pulse = 0, pulse = 1), or as an FSK signal. No ADC would be needed if digital data is being sent. The TDM receiver (*Figure 7-5*) picks up this signal, as well as the synchronizing pulses, and demodulates them. The receiver's output is sent through a Schmitt trigger, which reshapes the now ragged pulses, and into a demultiplexer. The demultiplexer places the binary data, located at its single input, into the appropriate output port for serial-to-parallel conversion — and on to digital-to-analog conversion (DAC) and filtering for output into a suitable transducer. Or, if the original baseband signal was digital data, D-to-A conversion is bypassed and the data is moved into either storage or on to a computer for immediate processing.

7-2. Data Communications

Pulse modulation entails the transmission of analog or digital modulation signals by varying either the width, amplitude, frequency, position, or phase of a pulse — or even whether a pulse is present or not. The pulses can then be decoded at the receiver. There are two basic methods to produce pulse modulation. The older *analog pulse modulation* (APM), and the more modern and more commonly used *digital pulse modulation* (DPM).

A method for transmitting binary information, as well as voice and video that has been digitized, at higher speeds than would normally be possible, is accomplished by employing *phase-shift keying* (PSK)

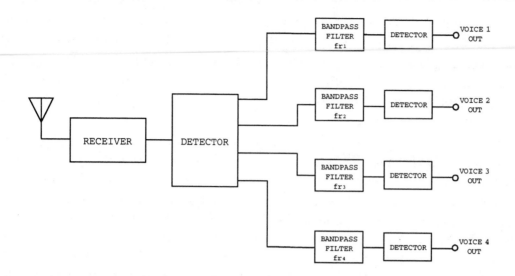

Figure 7-3. *A four channel FDM receiver system.*

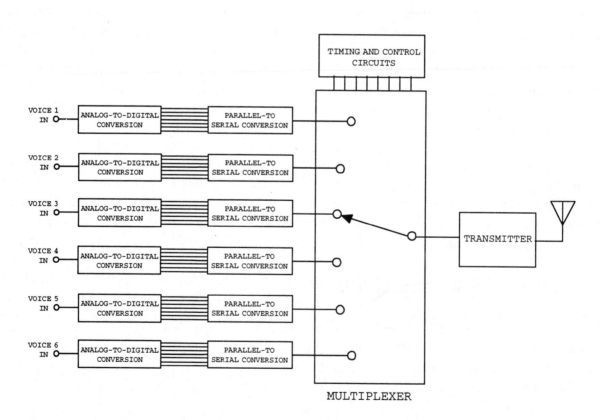

Figure 7-4. *A TDM voice transmitter showing multiplexer action.*

as well as phase shifting combined with amplitude modulation (*quadrature amplitude modulation*, or QAM). This allows the transfer of large quantities of information over space satellites or terrestrial wireless microwave links.

7-2A. Pulse Modulation

Analog pulse modulation permits the production of high values of peak power for short time periods, granting smaller transmitters higher output powers. A tiny sample of the modulating signal is taken, which is then allowed to continuously modify a certain characteristic of a pulse — since the carrier can be analog modulated by changing its width (PWM), amplitude (PAM), position (PPM), duration (PDM), or timing (PTM), then any of these methods can be adopted to impress pulsed information onto the RF carrier. Three of the most common of these techniques are shown in *Figure 7-6*.

With the help of pulse-width modulation methods, transistors can be adapted to the finals of some high powered AM amplifiers where, in the past, only vacuum tubes could have been used. By employing pulse-width modulation, transistors will be supplying current for something less than 100% of the time — but at the expense of a much more complex semiconductor circuit than that furnished by high powered vacuum tube finals. This class D amplification method transforms the audio signal into a pulse modulated signal train by using a pulse-width modulator. The output of this modulator is a rectangular signal that varies in pulse width and frequency to that of the changing modulating baseband signal. The transmitter's power amplifier is turned on and off by these pulses, allowing for a higher RF output due to the lowered duty cycle necessary. The PWM signal is then passed through tuned circuits and harmonic filters, which restore the original shape of the audio modulated carrier wave for transmission through the antenna and into space.

Pulse-coded modulation (PCM), a digital method for obtaining pulse modulation, has essentially replaced most of the above analog techniques for the transmission of voice, video, and data. Unlike the analog methods, PCM is virtually noise free and is able to punch through when atmospheric conditions are very poor, with little information loss. One way of obtaining PCM is by assigning a voltage level on the modulating audio waveform a specific digital value. (*Figure 7-7*.) For example, if the waveform of *Figure 7-7* were sampled at point four, the receiver would decode that amplitude as binary data 100. Or, if a signal has an amplitude of 3V and 5V at the sampling points, then a pulse train representing 011 and 101 would be outputted from the PCM modulator. In practice, many more sampling points must be used. For example, to send an 8 bit word, 256 sampling levels must be employed. A simple block diagram of a single-channel PCM digital RF communication system is shown in *Figure 7-8* and *Figure 7-9*. Either AM (CW or a single tone modulation) or FM (FSK) can be adopted as the medium for the PCM transmission and reception. *Figure 7-8* is the transmitter section of a non-multiplexed PCM/AM or PCM/FM system: the audio or (if the ADC is bypassed) the digital signals are transferred to a parallel-to-serial converter, then on to an AM or FM modulator to change the binary data into amplitude pulses or to frequency shifted "pulses". This signal is placed into the transmitter, which amplifies and filters the output for transmission from the antenna and into space. This signal is then received by the antenna of the PCM receiver of *Figure 7-9*. The PCM is amplified by the receiver, detected by the demodulator, and sent for pulse shaping by the *Schmitt trigger* — which squares the now ragged pulses.

Figure 7-5. *A TDM voice receiver capable of receiving six channels.*

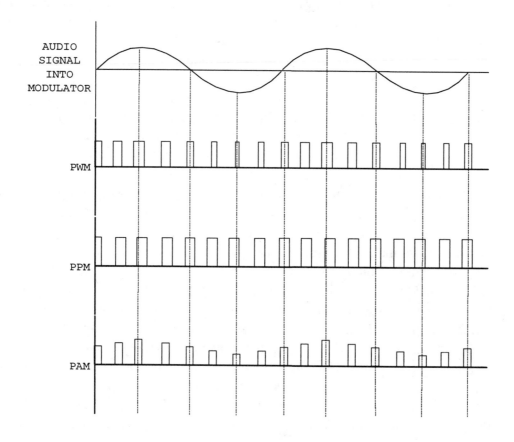

Figure 7-6. *Three of the most popular analog pulse modulation methods.*

These serial pulses are placed into the serial-to-parallel converter, which outputs parallel data into the digital-to-analog converter (DAC) for conversion into an analog signal. The DAC output is filtered by a low-pass filter to create an almost exact reproduction of the original audio signal.

7-2B. PSK and QAM Modulation

It is important to keep in mind that the wider the bandwidth of a channel, the more information that can be passed within that channel over a given time. A narrower communications channel means less information can be passed within a certain time frame — high amounts of data can still be sent through a narrow communications link, but the time it takes to pass this information will be longer. A way of helping to overcome some of these limitations is to utilize multilevel encoding schemes. In other words, each change of state of a digital signal can indicate two or more different binary values instead of the usual two-level binary (high = 1 and low = 0) arrangement. This standard two-level binary system only allows a change of state that indicates either a 1 or a 0, while with multilevel encoding each change of state may indicate a 00, a 01, a 10, or a 11, permitting twice the information to be transferred within the same bandwidth and time period. Most of these digital modulation methods have been developed for use in microwave digital radio high-capacity information transfers. A basic digital radio that employs some of these high speed processes is shown in *Figure 7-10*. The coder chip digitizes the incoming voice signal, which is sent to the modulator. The modulator can employ any type of PSK or QAM modulation method (explained below), which affects the phase of the carrier oscillator output or both its phase and amplitude. The output of the modulator is sent through a bandpass filter and into a mixer — with its accompanying transmit oscillator — for up-conversion. The resulting frequencies are then placed into the power amplifier and bandpass filter and out to the antenna for transmission through space. The receiver of *Figure 7-11* receives this signal, filters it through a BPF, then further amplifies it with the RF amplifier, which passes it into the down-conversion circuits. Down-conversion transforms the very high RF into a low IF of either 70 or 140 MHz, which is then filtered by the IF BPF and amplified by the IF amplifiers. The output of the IF amplifiers is sent for demodulation. The recovered signal is then fed into the decoder, which converts this digital input into a reconstructed analog output.

The above modulator can consist of one of many different PSK (phase-shift keying) techniques which, as explained above, changes the phase of the carrier to denote a digital "0" or "1". Either a 90° or 180° change of phase of the carrier signal marks the beginning of a digital "0" or "1." (*Figure 7-12*) Or, in modern BPSK (binary phase-shift keying), the constant amplitude carrier's phase is shifted between 0° and 180° for a "0" (0°) and a "1" (180°) in response to the coder chip's digital output. There are other methods, such as *DPSK* (differential phase-shift keying). DPSK is similar to BPSK but can employ simpler circuitry since DPSK does not need to reinsert an exact on-frequency carrier, which is furnished by the BPSK's elaborate carrier recovery circuitry, to supply the phase comparisons for proper data recovery. DPSK instead compares phase shifts from a received bit to an earlier received bit. *QPSK* (quadrature phase-shift keying) uses four different phase values such as 0°, 90°, 180°, and 270° on a constant amplitude carrier to furnish two bits for each change of phase (00, 01, 10, 11), instead of 1 bit (1, 0) as in the above PSK techniques. This method would obviously provide twice the information within the same bandwidth and time period. But perhaps the most popular modern method for transmitting data at very high rates utilizes a combination of amplitude and phase modulation, and is referred to

Figure 7-7. Conversion of an analog waveform into discrete voltage levels and PCM.

Figure 7-8. A basic single channel PCM transmitter.

Figure 7-9. A basic single channel PCM receiver.

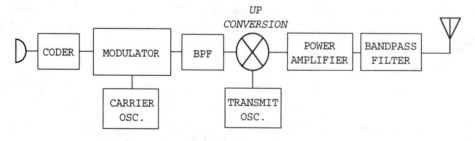

Figure 7-10. A digital radio transmitter.

as *QAM* (quadrature amplitude modulation). QAM is the major long distance digital communications technique employed for transmitting data over the air across terrestrial microwave links. QAM applies different phase shifts to the carrier, with each of these phase shifts having the capability to have two or more amplitudes. Each amplitude-phase combination can represent a different binary value. For instance, if a signal had a phase shift of 180° and an amplitude of +2, then it may represent a digital value of 111; if the phase is shifted to 90° with a amplitude of -1, then a 010 may be represented. This is referred to as 8-QAM, which uses four phase shifts and two carrier amplitudes for a total of eight possible states, or 3 bits (000, 001, 010, 011, 100, 101, 110, 111), that can be transmitted. 8-QAM is quite common, but is being supplanted by higher QAM values. As the number of possible AM/PM states increases, more data can be sent within an allotted bandwidth or time frame because more bits per change can now be encoded. *Figure 7-13* clarifies the basic idea of this type of modulation on what is referred to as a *constellation diagram*, in this case for a 16-QAM scheme (which supports 4 bits per AM/PM change). Constellation diagrams can display both amplitude and phase changes with their given binary coded values. However, as AM/PM states increase, noise starts to become more of a consideration, since each state becomes very close together. High QAM values can be subjected to crippling interference under relatively low noise conditions. This is why high noise environments, such as that encountered in satellite communications, usually adopt the simpler schemes, such as BPSK and QPSK. However, low noise terrestrial *radio links* using 16, 32, 64, 128, and even 256 QAM are now in operation. These newer methods can transfer data at far higher rates than older and simpler digital systems within the limited bandwidth constraints of most information channels.

7-3. Satellite Communications

Communication satellites are simply a repeater and antenna combination that amplify an incoming signal and retransmits this signal at an extremely high altitude (*Figure 7-14*), allowing for a very large area of coverage. Most are in geosynchronous orbit around the earth at an altitude of 22,300 miles, which allows the satellite to revolve once around the planet in a single twenty-four hour period. This permits the satellite to remain stationary over a chosen point above the earth, day in and day out. An earthbound antenna does not have to track a geosynchronous communications satellite as the earth revolves — it merely needs to be aimed once during installation and then locked in place. The extreme distance of the satellite from the receiver and transmitter, however, has a negative effect: The earth station antennas, transmitters, and receivers must be of high gain to effectively access these satellites to obtain a usable signal-to-noise ratio.

As shown in *Figure 7-14*, the earthbound transmitter sends a signal to the satellite (the *uplink*), where the satellite amplifies and translates (changes the frequency of) the signal, generally to a lower frequency, and retransmits this frequency (the *downlink*) to an earth station receiver. Without this *frequency translation* the very low-powered signal received at the satellite would be inundated by the far more powerful satellite transmitter's frequency. The transceiver of a satellite that performs this function of amplification and translation is referred to as a *transponder*.

All modern satellites utilize some form of multiplexing (either TDM or FDM) as well as multiple transponders set to different channels, allowing for the handling of huge amounts of information. Satellites

Figure 7-11. *A digital radio receiver.*

Figure 7-12. *A phase-shift keying waveform in response to the modulating binary data.*

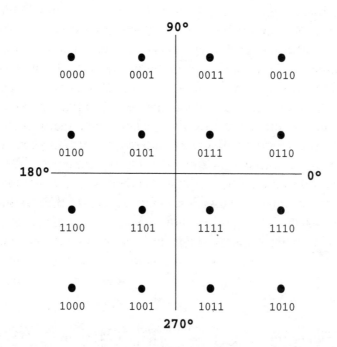

Figure 7-13. *A constellation diagram for 16-QAM.*

also exploit very wideband transponders, which permit any earth transmitter's signal within the transponder's bandwidth to be received and retransmitted. In view of this wide bandwidth, and the need for small, highly directional antennas, microwave frequencies are the only practical section of the RF spectrum exploited for satellites. These frequencies are normally in channels between 4 and 14 GHz.

7-3A. Frequency Reuse

Frequency reuse can double the amount of channels that a single satellite can retransmit within its band of allotted frequencies. A separate bank of transponders are employed, with one bank using a vertically polarized antenna, while the other bank uses a horizontally polarized antenna. Each bank utilizes the same frequencies as the other bank, but because a transmitting antenna of the opposite *polarization* as the receiving antenna cannot induce a voltage into that antenna, very little interference occurs. When *circularly* polarized antennas are operated for frequency reuse, they are wound in the opposite sense for each bank of transponders, inducing little or no voltage in the receiving antenna, thus also isolating each set of transponders. Another method, called *spot beaming*, uses a highly selective antenna to service two different stations, at separate locations on the earth, with the same set of frequencies.

7-3B. The Transponder

A simple transponder is shown in *Figure 7-15*. The uplink antenna receives the signal. This extremely feeble signal is then amplified by a low noise amplifier to obtain a decent S/N ratio, mixed with the LO frequency to acquire a sum and difference frequency, fed to another amplifier, then out to a bandpass filter (BPF) to achieve the proper frequency translation for the downlink channel. The downlink frequency's strength is increased by the power amplifier, which is typically a *traveling wave tube* (TWT). TWTs are not completely linear and produce *intermodulation distortion products*: the input signal mixes with any nearby signal or its harmonics, creating undesired sum and difference frequencies — which must be removed by another bandpass filter. This output is then fed into the downlink antenna for retransmission to an earth station. Other satellites use transponders with *double-conversion* circuits, or employ *regenerative repeater* transponders. (*Figure 7-16*.) Regenerative repeater transponders are transponders that are more like separate receiver-transmitters than transceivers. A signal is picked up by the receiver, amplified, converted, and demodulated. The output of the demodulator is a reproduction of the original digital, voice, or video signal transmitted by the earth station. This modulating signal is then further amplified and fed into the modulator of the transmitter section, which impresses the carrier with this baseband information. The modulated carrier is sent to the final power amplifier and BPF, and out to the downlink antenna. This scheme offers better signal-to-noise ratios, with increased and cheaper amplification of the uplink signal possible. (Since the low frequency baseband signal is far easier to amplify than a higher RF or IF of the single and double conversion types.)

7-4. Spread Spectrum

Spread spectrum (SS) is a communications system that spreads its desired signal randomly over a wide bandwidth of the frequency spectrum. Any receiver must be synchronized with the transmitter in order

Figure 7-14. *A communication satellite with its earth stations.*

Figure 7-15. *A single-conversion satellite transponder.*

Figure 7-16. *A regenerative repeater transponder.*

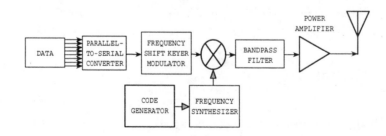

Figure 7-17. *A frequency hopping spread spectrum transmitter.*

to receive the transmitted signal. SS is most commonly used in data communications over wireless *LANs* (local area networks), as well as in industrial and military applications; and recently in consumer 900 MHz cordless phones and wireless biomedical monitors. SS is low noise, and resistant to jamming, eavesdropping, man-made interference (QRM), and certain types of fading. Band utilization is also maximized because the transmitter does not "own" any single frequency. Digital data as well as voice are both equally easy to implement in SS.

There are different ways to implement spread spectrum — with the two most prevalent being *frequency hopping* (FH) and *direct-sequence spread spectrum* (DSSS). Spread spectrum's exploitation for commercial operation is in three specific frequency bands between 902 and 5,850 MHz, and for the amateur services above 420 MHz.

7-4A. *Frequency Hopping*

The most common SS technique is referred to as *frequency hopping spread spectrum*, with the transmitter shown in *Figure 7-17*. The digital data to be transmitted feeds its parallel bits into a parallel-to-serial converter, which outputs the necessary serial binary pulse trains into a frequency-shift keyer (FSK) modulator, which converts this square wave binary data into two varying frequencies. (*Figure 7-18.*) One frequency denotes a binary "0", while the other frequency represents the binary "1." (FSK is an FM system with two frequency states — any frequency can be used for these two states, but are typically in the low audio range.) The FSK modulator outputs its stream of two tones, representing the serial binary data, into a mixer. A frequency synthesizer, usually a *PLL*, is driven by a code generator (or computer with the proper hardware and software) called a *pseudo-noise generator* (PN). The combination of the frequency synthesizer and the code generator performs the function of a *transmit oscillator* that is constantly changing its frequency output at preset times. The code generator supplies the PLL with divide-by-N counter control words. The output of the doubly-balanced mixer is the sum and difference frequencies of the signal and the transmit oscillator frequencies. The sum or difference frequencies can be filtered out and applied to a power amplifier and sent to the antenna for transmission.

The time that the code generator allows the frequency synthesizer to remain on any single frequency is called the *dwell time*, and is typically between 10 and 400 milliseconds (*Figure 7-19*), differing by design. Depending on the programming in the code generator, the carrier may jump to fifty or more frequencies before the original sequence is again repeated. A receiver must, of course, have the same programming in its code generator, as well as be in synchronization with the transmitter, to receive the transmitted signal. The receiver of *Figure 7-20* is capable of receiving this SS signal. A synchronizing code is received by the antenna, setting up the receiver to be locked in time-step with the transmitter. The receiver's LO, consisting of the frequency synthesizer, is now ordered to change frequency over time by the code generator, which keeps it in step with the frequency changes occurring within the transmitter. The output of the mixer is always at the fixed tuned IF. The IF section then amplifies this intermediate frequency, inserts it into the frequency-shift keyer demodulator, which outputs serial binary data, in square wave form. This output is usually sent to a serial-to-parallel converter for insertion into some type of processing equipment, such as a computer.

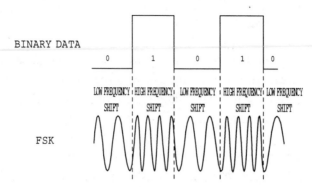

Figure 7-18. *An FSK waveform in response to modulating binary data.*

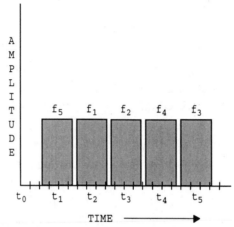

Figure 7-19. *Frequency shifting sequence in spread spectrum communications: Frequency f5 dwells at time t1, then shifts to frequency f1 at time t2, etc.*

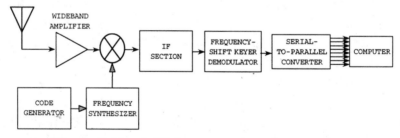

Figure 7-20. *A SS frequency hopping receiver.*

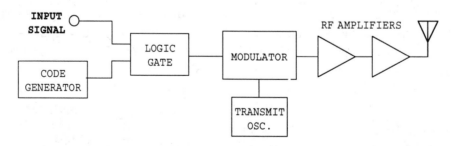

Figure 7-21. *A direct-sequence spread spectrum transmitter.*

7-4B. *Direct-Sequence*

The direct-sequence spread spectrum (DSSS. Also referred to as *code division multiple access*, or CDMA) method of producing spread spectrum is not as easily understood. One type of DSSS transmitter is shown in *Figure 7-21*. The modulating signal is placed into a logic gate (an XOR) along with the high frequency digital code generator output. The output signal from the logic gate is then fed into a modulator along with a transmit oscillator output. The modulator can produce any of the PSK or QAM modulation outputs. Since the modulator is commonly a doubly balanced mixer, the RF carrier is attenuated and the output now consists of various random wide-ranging sidebands, or as a "spreading" of the resulting signal. (*Figure 7-22*). This produces a very wide signal that appears to other standard receivers as noise. The DSSS receiver (*Figure 7-23*) picks up this wideband "noise", sends it to a doubly-balanced mixer, where it is combined with a comparable digital code produced by the code generator. The code generator supplies a comparable code as that generated by the transmitter. Synchronization of this code generator is maintained by sync circuits, which can strip-off special synchronizing clock pulses from the transmitter's signal (or by other complex methods). The resultant output of the mixer is now a "despread" signal, which is transferred to a regular receiver where it is demodulated and fed to a speaker — or to some digital device for further processing or storage.

7-5. Cellular Communications

Instead of using one large repeater to receive and retransmit radio signals, as in older mobile phone communication systems, *cellular radio* employs many multiple repeaters, or *cell sites*. (*Figure 7-24*.) Each cell site contains a low-powered transceiver designed to have limited range. This range overlaps with the coverage area of the next cell site, which operates at a different frequency. Frequencies are only repeated several cell sites down so that interference is minimized; but frequency availability and use is maximized. All of these cell sites are connected and controlled by the *MTSO* (*mobile telephone switching office*) through regular telephone lines, which also passes the cellular phone signal to the wired or wireless number called by the cell phone user. As a mobile user passes from one cell site into the coverage area of another, *hand-off* occurs. The cell site and the MTSO monitor the signal power of the user's cell phone. As the phone's signal begins to fade, the cell site sends a signal to the MTSO, which then orders the neighboring cell sites to monitor the signal strength of the user's phone. The cell site that has the strongest signal indication is ordered by the MTSO to pick up the user's phone signal, but on a different frequency. This instantly frees up the prior cell site for another call, and is generally not noticed by the cell phone user. Carrier frequencies are in the 800 MHz band, with over six hundred fully duplexed channels obtainable. When more cell sites are needed as the subscriber base increases, each cell site can lower its output power as more sites are installed, which increases the cell sites available for usage, while nullifying any possible interference between these sites.

7-5A. *The Cellular Transceiver*

A complete working cellular radio is shown in *Figure 7-25*. After the caller inputs a telephone number through the keypad and picks up the unit, the phone automatically attempts to institute communications

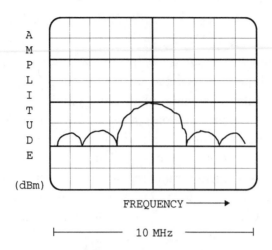

Figure 7-22. *A DSSS signal in the frequency-domain.*

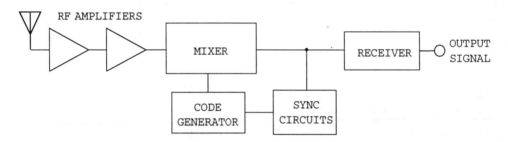

Figure 7-23. *A basic DSSS receiver.*

Figure 7-24. *Overlapping cell site coverage.*

with a cell site. After communication is established, the MTSO is automatically sent the ID number (which is the same as the phone number for the unit) for that particular cell phone. The user then presses the send key and the system finds a free channel. Information on what frequency and signal strength the handset should use is sent to the cell phone from the MTSO through the cell site. The MTSO then makes contact with the desired number as dialed by the cellular phone user. The user speaks into the microphone, where the voice level is raised by the audio amplifier. The frequency synthesizer has already been commanded by the MTSO through the logic control unit (LCU) to feed a certain carrier frequency into the transmitter section's mixer. Thus a channel has been effectively selected by the MTSO. This carrier frequency is placed into the phase modulator, where it is modulated by the audio and further amplified by the voltage and power amplifiers and inserted into the microstrip *directional coupler*. A directional coupler is simply a device that passes, in a single direction only, most of the microwave power that is placed at its input, while tapping off a small sampling of this input power to another of its ports. This tiny specimen of input power is fed to a power output sensor, were it is used by the APC (*automatic power control*). As well, the level of this power is sent to the MTSO, which can then order the APC circuit to decrease output power if it feels this energy is excessive, or to increase it as needed — up to a maximum of three watts. The directional coupler feeds most of the signal into a low-loss *duplexer*, which can consist of two notch filters or two bandpass filters. The duplexer allows the cell phone to use one antenna for both transmitting and receiving by isolating the receiver section from any of the transmitter section's RF power — but permits the transmitter to output its RF energy and the receiver to obtain an incoming signal from the same antenna. Since the transmit frequency is separated by 45 MHz from the receive frequency, the filters effectively notche or filter out the unwanted transmit frequency, preventing receiver overload and damage. When a signal is received by the cell phone, it proceeds from the antenna into the duplexer, which sends it into the RF amplifier and into the mixer. The first LO frequency has already been set by the MTSO through the logic control unit to the proper receive channel, as outputted from the frequency synthesizer at port LO1. This frequency mixes with the signal frequency, producing the first IF (either 82.2 or 45 MHz), which gets amplified and filtered. This IF is then inputted into the second mixer to obtain the second IF (either 10.7 or 0.455 MHz). This IF is detected by the FM demodulator to recover the voice modulation, then amplified and sent to the speaker, while the bandpass circuit filters this demodulated output to recover the control signals for the logic control unit. These control signals were generated by the MTSO. The RSSI (*receive signal strength indicator*) amplifier feeds the strength of the received signal, as a DC level, into the logic control unit, which outputs the proper indicator signals to the MTSO through the transmitter section to keep the MTSO updated. The MTSO can then, as needed, switch the cell phone to another cell site if signal strength is lagging. The microprocessor section controls the 12-key DTMF tones (*dual-tone multiple frequency*) for dialing out, drives the LCD or LED display, stores the last number dialed and frequently called numbers, and other common functions.

7-6. CW Communications

CW (*continuous wave*, or "A1") communications equipment is still extensively used in amateur radio as well as mandated aboard certain oceangoing shipping vessels. A basic knowledge of their operation is certainly required of any RF technician.

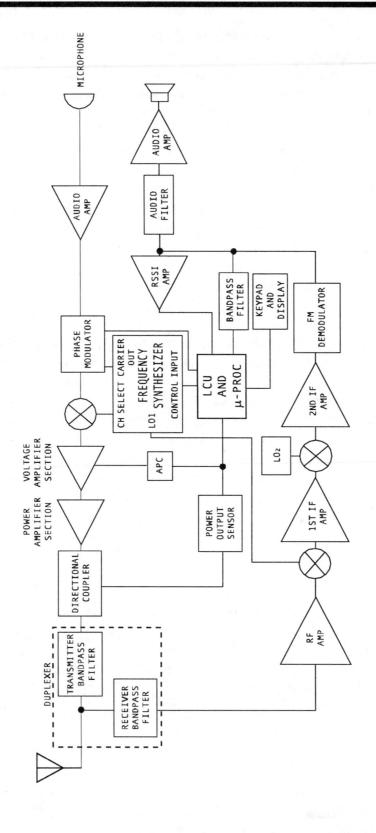

Figure 7-25. *A cellular radio transceiver.*

There are two basic methods used to transmit Morse code. One entails employing amplitude modulation and impressing the carrier with a 1000 Hz tone for reception by any receiver (*modulated continuous wave*, or MCW). This technique is rarely used today. The dominate method is interrupted CW, which involves switching on and off the carrier of the transmitter to form the dots and dashes necessary to transmit the desired information. (*Figure 7-26.*)

All amplifiers can be biased for highly efficient class C operation in a CW transmitter. Thus, little power is consumed unless an actual dash or dot is being sent. However, linear stages are often used so that SSB can be transmitted, as well as the higher quality code output inherent with linear stages. Also, since there is little concern about maintaining the various frequency's phase relationships, as in AM and FM transmissions, there is less fading of the received signal, while there is significantly lower noise received due to the very narrow bandwidth of a CW signal. Moreover, a single tone is far easier to read than the complex waveforms of speech under poor receiving conditions.

7-6A. CW Receivers

The standard CW receiver (*Figure 7-27*) is of the superheterodyne type, and varies little from AM, SSB, or FM receivers. Nevertheless, in order to properly receive CW, a receiver must incorporate a BFO (beat-frequency oscillator) — in fact, because all SSB receivers have a built-in beat-frequency oscillator, they are automatically able to receive high quality code. Code can even be heard through a simple AM diode detector receiver, but the signal is a very mushy, hard to read humming sound. To obtain its signal, a CW receiver's BFO mixes with the IF in a *product detector*, which is basically a nonlinear mixer. The BFO is set to almost the same frequency as the incoming IF, but is offset by a prefixed frequency, usually about 800 Hz. This offset means that the output of the mixer will be a 800 Hz sine wave tone — or the difference between the BFO's frequency and the intermediate frequency. The IF, the BFO frequency, and the sum frequency are too high to pass through or to be amplified by the audio amplifiers.

7-6B. CW Transmitters

A CW transmitter is basically an amplified carrier oscillator. A method is needed to switch on and off this carrier, usually by removing or supplying a path for the DC power to a stage or stages — either directly or through a relay or semiconductor switch (a "keying transistor"). Many keying methods can be utilized. Some allow the operator, when employing a straight telegraph key, semiautomatic or fully automatic "bug", or even a computer, to interrupt the emitter or cathode return of the final amplifier. (*Figure 7-28.*) As soon as the operator depresses the sending key, the DC return is reestablished, allowing the stage to become operational and transmitting the carrier into space. All of the emitters or cathodes can be tied together to switch the entire oscillator and amplifier chain on and off (*Figure 7-29*) — or just the oscillator itself can be turned on and off. Additionally, the Vcc to the collector or plate of the amplifiers can be controlled (*Figure 7-30*) by either removing or supplying the DC power. A very common keying method is called *blocked-grid keying*. (*Figure 7-31.*) This procedure supplies a bias that cuts-off the transistor's or tube's base or grid, allowing no collector or plate current to flow until the Morse key is depressed, shorting out the blocking voltage to ground and allowing the stage's normal bias to permit amplification of the transmitter's CW signal.

Figure 7-26. *An RF carrier turned on and off to transmit information in Morse code.*

Figure 7-27. *A basic superheterodyne CW receiver.*

Figure 7-28. *Keying a CW transmitter by breaking and reestablishing the emitter or cathode return of an amplifier.*

Figure 7-29. *Keying a transmitter by controlling the DC cathode return of the oscillator and final amplifier.*

Figure 7-30. *Keying a transmitter by switching on and off an amplifier's collector or plate voltage.*

Figure 7-32. *CW waveform with a rapid rise and fall. Will cause key clicks.*

Figure 7-31. *Blocked-grid keying for a CW transmitter.*

Figure 7-33. *Filtered CW transmission. No key clicks.*

7-6C. CW Spurious Responses

If the carrier is too rapidly switched on or off (*Figure 7-32*), the fast rise and fall times created will cause side frequencies to be generated, producing what is known as *key clicks*, which can be heard by the receiver of the code. This will also cause unacceptable adjacent channel interference. If the rise and fall times at make and break are made softer, or extended, as in *Figure 7-33*, then the production of spurious radiation will be eliminated. This is usually accomplished by placing a capacitor across and a small choke in series with the keying device.

Chirp is a condition in a CW transmitter caused by the transmitter's oscillator being pulled off frequency when keyed (FMing), creating a birdlike chirping sound and adjacent channel interference. This is generally produced by poor regulation, a buffer or oscillator component failure, poor design, or other power supply problems pulling down the oscillator when keyed.

7-6D. A Solid-State CW Transceiver

Figure 7-34 shows the basic parts, in block diagram form, of a simple CW transceiver; although nowadays these circuits, or similar circuits, are ordinarily part of a complete SSB transceiver and share many of their common circuits. (See *Circuit Sharing*.) Furthermore, almost exclusive use of frequency synthesis for variable frequency tuning applications is predominate in the modern CW and SSB transceiver.

Looking at the block diagram of *Figure 7-34*, when the T/R switch is set to *transmit*, a relay-driving transistor energizes an electromagnetic relay. This relay switches the antenna into the transmitter circuit and out of the receiver circuit while also supplying power to the transmit section — this too places an offset capacitor in the VFO circuit to provide a transmitter offset of approximately 800 Hz by slightly varying the capacitance of the VFO's frequency setting components. The input to the audio preamp is also grounded through a muting transistor during this period to provide the necessary receiver muting. When the telegraph key is then closed for transmission of a dot or a dash, the driver's collector is supplied with its Vcc through a switching transistor, turning it on. An 800 Hz tone is also produce by the sidetone generator for use by the operator, and amplified by the audio power amplifier to drive the speaker. The sidetone indicates to the operator when the telegraph key has been depressed and aids in the sending of code. The VFO's output is fed to a buffer amplifier, to isolate the oscillator from any possible pulling effect of subsequent stages, and sent into the balanced mixer. The buffered HFO (*heterodyne frequency oscillator*) frequency is injected into the other input port of the balanced mixer. The HFO selects the desired band through the operator manually switching in frequency-determining crystals. The VFO is mixed with this crystal controlled HFO frequency. The output is amplified and filtered by the mixer amplifier to obtain the VFO minus the HFO — or the difference frequency. With this type of *pre-mixing* scheme, any frequency within a certain range can be tuned by the air-dielectric *tuning capacitor* for an almost drift-free output from the LC VFO. The signal is sent through the driver, filtered, and transferred into the final amplifier and out through an adjustable LC filter and matching network and into the antenna.

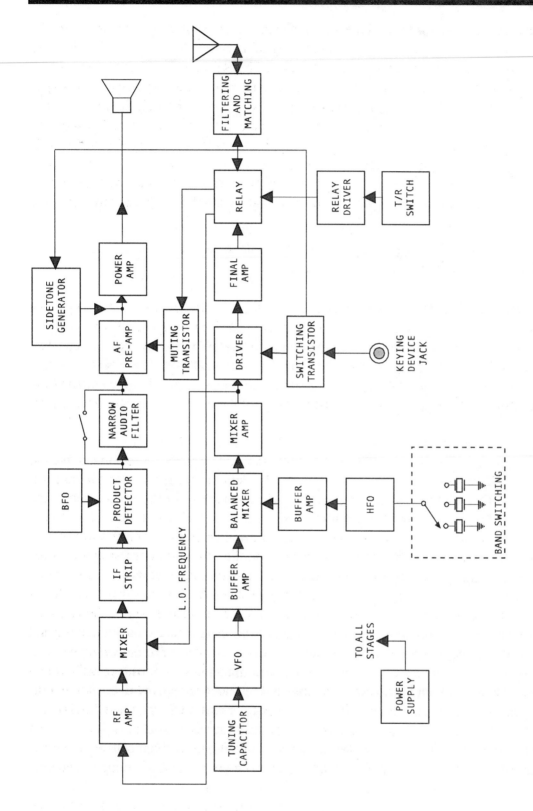

Figure 7-34. A complete multiband CW transceiver.

When the T/R switch is set to *receive*, the receiver section is unmuted, the antenna is switched into the receiver and out of the transmitter circuit. Any incoming signal is now filtered by the filtering and matching network and amplified and further filtered by the tuned FET RF amplifier. It is then sent to the mixer to heterodyne with the local oscillator frequency from the mixer amplifier output, which produces a sum and difference frequency. The difference frequency is able to pass through the tuned output circuit of the mixer where it is amplified by the IF strip, which supplies most of the receiver's selectivity and amplification. The IF is then coupled to an IC balanced product detector. A crystal-steady BFO mixes with the IF, creating an audio tone difference, which is sent to the active narrow audio filter to restrict the received bandwidth if interference and noise is a problem. If not, the narrow filter can be switched out. The tone is transferred to the audio preamp for voltage amplification, and then into the final amplifier to power the speaker.

7-7. Repeaters

Repeaters are used to extend line-of-sight communications over long ranges. (*Figure 7-35.*) Repeaters have been operated by police, fire, the amateur services, and local trucking companies for years, typically in FM. Long-haul, high capacity telephone transmission by complex microwave link repeaters utilizing multiplexing and exotic modulation schemes is also commonly practiced. The repeater concept takes advantage of increased antenna height and re-amplification to extend a small low-power and low-lying transmitter's communications range beyond the curvature of the earth or over any obstruction, such as mountains.

One type of repeater is shown in block diagram form in *Figure 7-36*, and displays the antenna, duplexer, receiver, controller, and transmitter sections common to most repeaters. The repeater user will transmit on the repeater's *input frequency*, which is the frequency of the repeater's receiver section, and, if the repeater so requires, a *CTCSS* or a *DTMF* tone, riding on the user's transmitted carrier, relays to the repeater's *controller* an access code to obtain repeater access and switch on the repeater's transmitter section. If no access code is required by that repeater, then the controller is replaced by a simple *carrier-operated relay* (COR), which will allow access to the repeater by any transmitter sending on the proper frequency with a sufficiently powerful signal. The COR is activated when the receiver's squelch is overridden, and will then open the repeater's transmitter section for retransmission of the received signal. The repeater's dedicated receiving antenna, or a single common antenna for both receiving and transmitting, can be employed to receive a radio operator's signal. The single common antenna must operate with a *duplexer*, which permits the receiver section to receive signals while allowing the transmitter section to send signals, with both the receive and transmit frequencies separated, or *split*. (Common split frequency spacings between the input and output frequencies are 600 kHz and 5 MHz, depending on carrier frequency.) This eliminates the massive interference that would result by using a single common receive and transmit frequency. Retransmission of the received signal is at the *output frequency* into a transmitting antenna or into the duplexer for transmission out of the common antenna.

A typical twin-antenna repeater must employ high-Q tuned cavities (see *Cavity Resonators*) in the separate transmitter and receiver feedlines to decrease any receiver desensitization created by the simultaneous operation of the two units in such close proximity.

Figure 7-35. *Repeaters are typically placed on high ground to extend communication ranges.*

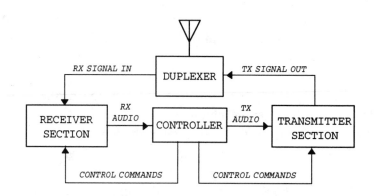

Figure 7-36. *A single antenna repeater.*

Figure 7-37. *A radiolink system.*

7-7A. *Radiolink*

As stated above, transmission of telephone conversations from coast-to-coast can be implemented with digital microwave multiplexed relay (repeater) stations, referred to as *radiolink*, and the newer ISDN (*integrated services digital networks*) systems, with each link able to carry thousands of simultaneous communication channels at up to 155 Mbits per second using special modulation techniques. (However, not all radiolinks have been completely digitized as yet.)

Radiolink microwave repeaters must be placed every 20 to 40 miles on tall towers using high-gain dish antennas. (*Figure 7-37.*) Radiolink is adopted instead of high-capacity optical fiber cable if there are no continuously available installation routes for the cable; or if bad terrain would make installation too expensive or difficult; or when quick installation of the communication's link is a priority.

To transmit huge amounts of data, even with time-division multiplexing techniques, consumes significant bandwidth. This demands the use of the microwave bands, commonly of 4, 6, 7, 8, 11, 14, or 18 GHz. (Short range intercity links are being implemented in bands between 15 and 50 GHz that provide high capacity transmission capabilities.) Above 18 GHz, rain attenuation of a long-haul microwave signal becomes a significant problem.

Different radiolink systems apply different modulation methods, such as AM, FM, PM — or the most popular modern technique of all, AM/PM, referred to as *QAM*. (See *Digital Pulse Modulation*.)

CHAPTER 8
Microwaves

Microwave communications is utilized mainly by telephone companies employing multiplexed micro-wave terrestrial and space-located repeater stations (See *Radiolink* and *Satellite Communications*), instead of copper or fiber-optic cable. As well, certain electronic television news gathering (ENG) and amateur services applications employ microwaves for their communications needs, while modern satel-lite cable TV delivery need not be made to a single distribution point as in the past, but also directly to the cable subscriber's personal receiving dish (satellite TV).

Microwaves are in the high-end of the radio frequency spectrum between 300 MHz and 300 GHz. Normal RF amplifiers cannot be used in this microwave region because of an increase in degenerative feedback created by the interelectrode capacitance and *transit time* problems of the transistors and vacuum tubes. This causes the gain of the amplifier, as the frequency increases, to become poor. (See *Amplifier Frequency Limitations*.) Also, because the values of any reactive components must become smaller and smaller as the frequency is raised, their physical dimensions in any amplifier circuit reach a point where they are no longer viable. In addition, stray reactances of even a single piece of wire or the reactance of a common resistor can adversely affect circuit operation. Special components, certain wiring considerations, and different designs for amplifiers must therefore be implemented.

8-1. Microwave Semiconductors

Microwave semiconductor amplifiers and oscillators are reaching higher and higher frequencies and are displacing the once dominant vacuum tube in low-power microwave applications. With the utilization of exotic materials and designs, microwave transistors are ever increasing in power output, frequency range, and importance in a field that was, until recently, dominated by the traveling-wave and klystron tubes. Furthermore, for some years now, microwave diodes have been able to operate at high micro-wave frequencies as oscillators, mixers, switches, and detectors.

8-1A. Microwave Diodes

Schottky barrier diodes (also called *hot-carrier*, or HCDs) are used as microwave detectors, harmonic generators, and mixers. The Schottky diodes are mechanically strong and have a low forward barrier voltage, enabling reception of low signal levels at up to 100 GHz.

The *point-contact* diode is able to perform as a microwave detector and mixer. A very fine wire (a *cat's whisker*) is utilized as the cathode, which makes contact with the silicon anode chip. (*Figure 8-1.*) This point of contact has extremely low capacitance due to the small surface area involved. The diode needs only a low forward voltage, which makes it perfect for low signal applications. It is, however, very fragile physically, and cannot survive large signal currents or mechanical impacts.

Figure 8-1. A microwave point-contact diode showing its internal structure.

Figure 8-2. An harmonic generator for frequency multiplication.

Figure 8-3. A Gunn oscillator with tuning-screw frequency adjustment.

Figure 8-4. A Gunn oscillator with YIG frequency adjustment.

Figure 8-5. A Gunn oscillator with varactor frequency adjustment.

Figure 8-6. A high frequency junction-tetrode transistor.

Gallium arsenide *varactor diodes* are utilized in the microwave region, commonly in frequency multipliers, as are *step-recovery* ("snap") diodes. In the frequency multiplier circuit of *Figure 8-2*, multiplication is obtained as the signal at the input turns on and off the varactor, or a step-recovery diode, creating numerous harmonics due to the diode's nonlinear response to this input frequency. The circuit does not contain discrete components as a lower frequency circuit might, but is instead etched on a circuit board in the form of microstrip or stripline, with the bandpass filter for the output frequency and the bandstop filter for the second harmonic commonly being cavity resonators. Unlike in an active frequency multiplier, there is an insertion loss — however, this circuit requires no external power source to operate.

PIN diodes are operated in the microwave region as current controlled resistors used to attenuate an incoming signal, as AM modulators (see *PIN modulator*), or to switch signal paths between circuits. Attenuation of an incoming signal is varied depending on the amount of forward DC bias placed across the diode. More forward bias decreases the resistance, while a decrease in forward bias increases the diode's resistance. Signal switching works by supplying a forward bias to the diode, which turns on the device, permitting a signal to cross along with the DC bias. A reverse DC bias turns off the device, blocking the signal.

Gunn diodes are used as microwave oscillators, with their frequency of oscillation dependent on the transit time of the electron through the device and, when placed in the proper resonant cavity, the diode can be made to oscillate at up to approximately 50 GHz. Its output power is contingent on the diode's design frequency — the thinner the semiconductor, the higher the frequency, but the lower the output power, while the lower this operating frequency, the higher the output power possible (up to 3 to 4 watts). The Gunn device has a very limited frequency adjustment range when placed in a cavity with a tuning-screw (*Figure 8-3*), which varies the cavity's volume. For much wider frequency variations a *YIG crystal sphere* (yttrium-iron-garnet, *Figure 8-4*) is utilized, which tunes the frequency of the cavity in proportion to YIG coil current through an electromagnet. The bias across a *varactor tuning diode* (*Figure 8-5*) can also be changed to slightly modify the output frequency of the Gunn oscillator.

Other diodes are available for use in oscillators, such as *impatt* diodes, capable of high output powers (25 watts), but at the deficit of high noise generation; *trapatt* diodes, employed usually in high power pulse-mode only; and *barritt* diodes, which are low noise, but at low power levels of a few milliwatts. *Tunnel diodes* can be adopted as a high speed switch or made to oscillate at microwave frequencies. However, they have a low power output and become unstable with temperature changes. Tunnel diode use has declined over the years since their introduction, with the Gunn diode assuming most of their functions.

8-1B. Microwave Transistors

Gallium arsenide FETs (GASFETs, sometimes referred to as MESFETs) are more popular than microwave bipolar transistors, and must be used instead of the transistor if amplification above 10 GHz is needed — they are capable of amplifying up to 20 GHz. Also, FETs have higher gain, increased input resistance, better thermal stability, and lower noise than bipolar microwave transistors. The power MESFET is a high power GASFET with an output up to 50 watts at 1 GHz and several watts at 15 GHz.

Figure 8-7. A two-cavity klystron microwave tube.

Figure 8-8. Operation and internal structure of a magnetron.

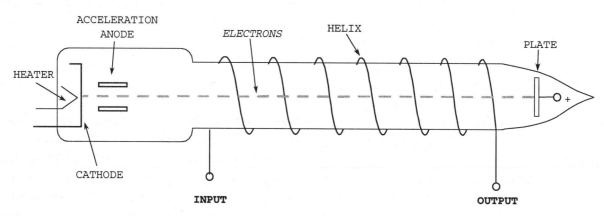

Figure 8-9. Operation and internal structure of a TWT.

Figure 8-10. Structure of microstrip.

Microwave bipolar transistors are operated in amplifiers and oscillators, are made only as the silicon NPN type, and are usually placed on $50\,\Omega$ microstrip. Maximum frequency is around 12 GHz, but they are generally used at 8 GHz and below. Most high frequency transistors have four leads, with one such type being the *junction-tetrode transistor* of *Figure 8-6*. This is a high frequency transistor with two base leads. The extra base terminal is supplied with 0.4V, which decreases capacitance and base resistance and thus increases high frequency response. Other four lead transistors can have double emitters, instead of the two bases. Older high frequency transistors, such as the *point-contact transistor*, being basically constructed with a base of silicon and a collector and emitter made from two fine wires, are now rarely used.

8-2. Microwave Tubes

For higher power and frequencies that can be obtained from semiconductors, vacuum tubes must be employed. Microwave tubes adopt the principle of velocity modulation to amplify a signal: Alternately increasing and decreasing the speed of a stream of electrons causes the electrons, when their velocity is decreased, to give up energy to an output probe.

8-2A. Klystrons

Twin cavity and multicavity klystrons are used as microwave amplifiers or oscillators. *Figure 8-7* shows a two-cavity klystron. A cathode inside the electron gun is heated, releasing a stream of electrons toward the positively charged beam collector plate. This beam of electrons must pass through two cavity resonators, which set the resonant frequency of the klystron. Frequencies of operation can commonly be between 3 to 100 GHz. The first cavity (the "buncher") is the input; the second (the "catcher") is where the output is tapped off. As the electron beam passes through the first cavity the input signal needing amplification velocity modulates this stream, which then excites the second cavity into oscillation at the cavity's resonant frequency. This power is tapped off by a conductive loop. Thus the original input's signal frequency causes the cathode's direct current electron beam to be transformed into an alternating stream of electrons, causing amplification of this low power input signal. Adding more cavities (*multicavity klystrons*) increases the amplifications of the input signal and can be used to increase its relatively narrow bandwidth.

Klystrons are capable of output powers of thousands of watts with tubes that can reach up to five feet in length.

For the klystron to function as an oscillator, some of its output signal is injected inphase into its input port, creating positive, or regenerative, feedback — and thus oscillations.

The *reflex klystron* is a single cavity klystron used as a low-powered microwave oscillator. It is normally found only in older equipment, with *Gunn diodes* being the replacement device in modern circuits.

Figure 8-11. *Structure of stripline.*

Figure 8-12. *A popular stripline high-Q bandpass filter used at microwave frequencies.*

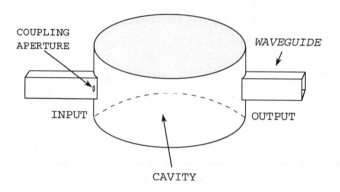

Figure 8-13. *A waveguide-driven cavity resonator.*

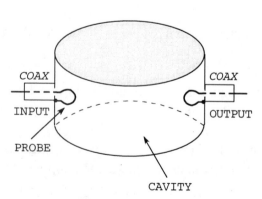

Figure 8-14. *A coaxial-driven cavity resonator.*

Figure 8-15. *A tunable cavity resonator.*

8-2B. Magnetrons

The *magnetron* is a high-powered, highly efficient narrow bandwidth microwave vacuum tube oscillator, usually found in radar transmitters and in some microwave transmitters. The magnetron consists of a unidirectional device (basically a diode) in which the magnetic field of a very powerful magnet passes. (*Figure 8-8.*) This magnetic field forces electrons to escape from the heated cathode (see *Thermionic Emission*), causing them to travel in a spiral path around the rod shaped cathode, which is surrounded by the positively charged anode. Due to the varying magnetic and electric fields affecting the electrons, they are speeded up and slowed down, releasing energy into the resonant cavities, causing oscillations. Some of this power is removed by a wire loop probe placed in one of the cavities.

8-2C. Traveling-Wave Tubes

Traveling-wave tubes (TWT) are used as amplifiers or oscillators up to a few hundred GHz. They are radiation hardened (for satellite use), very reliable, have a wide bandwidth and high gain, and can generate up to 20,000 watts of power. It is the premier microwave communication tube in use today.

Electrons are emitted from the heated cathode (*Figure 8-9*), into the acceleration anode, which accelerates and focuses the electron stream towards the positively charged collector plate. The input signal to be amplified is now injected into the helix. The helix slows down this signal to the speed of the electron beam, causing their two fields to interact, producing velocity modulation. More electrons are slowed down than speed up, which induces inphase voltages into the helix, causing amplification of the original signal. The output is taken from the opposite end of the helix. For use as an oscillator, some of its output is injected inphase into its input, creating positive, or regenerative, feedback.

8-3. Microstrip, Stripline and Cavity Resonators

Microstrip and *stripline* are both operated as short transmission lines and as tuned circuits and high-Q filters, while high frequency components (transistors, FETs, diodes) can be mounted directly on the microstrip and stripline's PCB (printed circuit board).

Cavity resonators are found, instead of conventional LC tuned circuits, at microwave frequencies, and are employed as bandpass or bandstop filters, or to set the frequency of an oscillating active device.

8-3A. Microstrip and Stripline

Microstrip (*Figure 8-10*) is usually one-quarter or one-half of a wavelength long, and is unbalanced. Due to its unshielded construction it can radiate RF. Balanced stripline (*Figure 8-11*), because of its twin ground planes, does not. Both commonly have a dielectric made of a printed circuit board of fiberglass, polystyrene, or Teflon©, while the conductor and ground plane are commonly made of PCB copper foil strips. Microstrip is cheaper and easier to fabricate since it can use standard PCB manufacturing techniques. Their characteristic impedances are ordinarily between 10 to 100Ω. As stated above, microwave components can be soldered to the printed circuit board along with the PCB's integral printed microstrip or stripline. The *stripline high-Q filter* (*Figure 8-12*), a common microwave bandpass filter, is also easily created by using a small tuning capacitor and copper traces on a PCB and enclosing the entire assembly in a shielded case. It has low insertion loss (around 1 dB), a narrow bandwidth, and steep skirts. Extra selectivity can be obtained by coupling together two or more such stripline filters.

8-3B. Cavity Resonators

A device called a *cavity resonator* can be used as a tuned circuit at microwave frequencies. The resonator consists of a hollow metal cavity that has the ability to select its own natural resonant frequency and reject all others. The volume of the cavity dictates its natural resonant frequency. Due to a very high Q, it has a very sharp filter response. A resonant cavity can be operated as a tunable bandpass filter driven by a waveguide through coupling apertures (*Figure 8-13*), or coax driven by 1/4 wavelength probes (*Figure 8-14*); or as the frequency controlling element in some microwave oscillators. They are commonly employed in repeaters to lower *desensitization* (a close unwanted frequency can cause an overloading in the first RF amplifier or the 1st mixer of a receiver, producing a loss of sensitivity to the desired signal), and can be placed in both the repeater's transmitter and receiver feedlines. A *duplexer*, a device that allows one antenna to be used for transmitting and receiving, can also utilize resonant cavities to pass the incoming signal from the antenna to the receiver, while allowing the transmitter to place its signal on to the antenna for transmission. Duplexers employ two cavities with each resonant to a different frequency — one frequency for the received signal, the other for the transmitted signal.

Cavity resonators can also be made tunable by mechanically changing their volume, as in *Figure 8-15*. As the tuning knob is rotated in one direction the flat disk will lower, decreasing the volume of the cavity, thus increasing the operating frequency of the resonator. As it is rotated in the opposite direction the volume is increased, thus lowering its natural resonant frequency.

CHAPTER 9
Antennas and Transmission Lines

Antennas convert alternating current into electromagnetic waves and inject these waves into free space; or receive electromagnetic signals from free space and convert them into an alternating current. There are many types of antennas available, and their design depends on the transmitting or receiving frequency, the directivity required, cost considerations, and space limitations.

Transmission lines are used to transfer this alternating current to or from a receiver or transmitter with minimal loss and maximum rejection of any external interference.

9-1. Antennas

A resonant antenna acts as a series resonant circuit. When an antenna is cut to a 1/4 or 1/2 wavelength, or typically multiples of a 1/2 wavelength, then maximum current flows through the antenna elements, thus giving maximum signal strength. (*Figure 9-1.*) Radiation patterns, however, vary with the transmitter's harmonic frequency when the antenna length remains the same. In *Figure 9-2a*, a dipole is being run on its fundamental frequency as a normal half-wave antenna; *Figure 9-2b* shows the same antenna operated on its second harmonic as a full-wave antenna; *Figure 9-2c* demonstrates the dipole running on its third harmonic as a one and a half-wave antenna. These changes in the radiation pattern of the antenna will obviously affect its directional characteristics, sometimes drastically. However, an antenna operated at its fundamental as a half-wave aerial, and at its third harmonic as a one and a half-wave antenna, will give them both the same low input impedance at their center-fed feedpoint, making harmonic operation possible with a single antenna. A proper impedance match between the transmitter, the feedline, and the antenna, will prevent the RF energy generated in the transmitter from being reflected back into its finals, possibly causing severe power output loss, transmitter damage, and transmission line insulation breakdown. A perfect match between the transmitter, its feedline and the antenna occurs when the inductive reactances equals the capacitive reactances and cancel (or $X_L = X_C$), and the pure resistances equal, typically, between 50 and 75Ω. (*Figure 9-3.*) When this point of matched impedances occurs, maximum power will be radiated from the antenna. To put it another way, the antenna must have an equal impedance match to its transmission line's *surge impedance*, or standing waves will result on the mismatched transmission line, and much of the subsequent reflected power will be given up as heat in the line. Surge impedance is the transmission line's natural input and output impedance, which is a function of its physical construction. When the antenna is resonant and the impedances between the transmitter, feedline, and antenna match, maximum alternating current is produced when the RF signal is injected into the elements of the antenna.

SERIES RESONANT CIRCUIT AT fr   SERIES RESONANT CIRCUIT AT fr

MAXIMUM CURRENT   MAXIMUM CURRENT

Figure 9-1. *Equivalent dipole antenna circuit.*

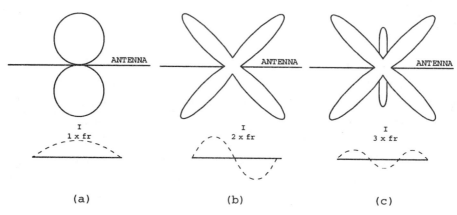

(a)   (b)   (c)

Figure 9-2. *Radiation patterns of dipole antenna run at its fundamental half-wave (a); as a second harmonic full-wave antenna (b); as a third harmonic 1 1/2 wave antenna.*

Figure 9-3. *A perfect impedance match between the transmitter, feedline, and antenna, resulting in maximum power output.*

Figure 9-4. *Current and voltage distribution on a half-wave resonant dipole.*

Figure 9-5. *A horizontally polarized antenna showing the horizontal electric fields.*

Figure 9-6. *A vertically polarized antenna showing the vertical electric fields.*

On the *antenna*, maximum standing waves produce maximum voltage at the ends of the antenna, and maximum current in the middle, as demonstrated using the dipole antenna of *Figure 9-4*. This creates maximum antenna output power. Radiation from the antenna is generated when the frequency reaches a point where the field lines produced by the high frequency alternating current cannot collapse back into the antenna before the alternating current changes polarity, which usually occurs at around 30 kHz. Some of this energy is radiated into space as an electromagnetic (RF) wave at the speed of light. The receiving antenna's elements are then cut by this wave, inducing a small voltage, which must be further amplified and filtered by the receiver's circuitry.

9-1A. Polarization

Both sending and receiving antennas should be properly polarized. Polarization refers to the orientation of the electric field of the electromagnetic wave through space. A *horizontally* polarized antenna (*Figure 9-5*, the antenna's elements are parallel with the ground) will receive an electromagnetic wave from a *vertically* polarized antenna (*Figure 9-6*, the antenna's elements are at right angles to the ground) only due to a small shift in polarization that occurs over distance. If it were not for this wave change, a vertical antenna could not induce a voltage into a horizontal antenna, and vice-versa. A way to overcome this problem is, of course, to make sure both antennas are appropriately polarized or, in the higher frequency range, the use of a *circular* polarized antenna, such as the *helical* type of *Figure 9-7*, will induce a voltage into either horizontally or vertically polarized antennas with very little loss (about 3 dB). However, when communicating with other helicals, they must be of the same *sense* (direction of winding), or extreme signal attenuation will take place between the transmitting and receiving antennas.

9-1B. Common Antenna Types

In general, a vertical antenna radiates in all directions (*Figure 9-8*, *omnidirectional*), and is commonly a quarter of a wavelength long, with the earth providing the other necessary quarter wavelength through ground reflection. (*Figure 9-9*.) This is the type regularly adopted for AM, FM, and TV commercial broadcasts, with the vertical tower itself acting as the 1/4 wave antenna, supported by non-resonant *guy wires*. Resonant guy wires would act as antennas, radiating the signal or its harmonics and increasing RF losses. The guys are broken up by *strain insulators* placed at the appropriate non-resonant sections of these wires.

Since the 1/4 wave vertical depends on a proper ground for completion of the other missing 1/4 wave, a high quality ground is essential. This can be a problem in some areas. The use of four or more 1/4 wavelength wire *radials* placed at the base of the antenna and parallel to the earth can be adopted as an artificial ground. This group of radials is called *counterpoise* or *groundplane*.

Horizontal antenna types have more directional characteristics — all the way from the bidirectional half-wave *dipole*, with the horizontally polarized radiation pattern of *Figure 9-2a*, to the highly directional *parasitic* multi-element *Yagi* radiation pattern of *Figure 9-10*. A common Yagi antenna is shown in *Figure 9-11*. Parasitic refers to an element, made of either wire or tubing, that is not directly driven by the feed line, but has a voltage induced into it by the *driven* element. A single driven element, a dipole,

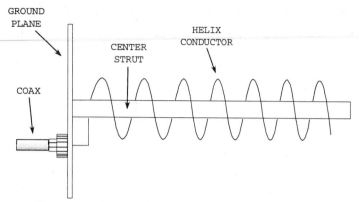

Figure 9-7. *A high frequency circularly polarized antenna of the helical type.*

Figure 9-8. *Radiation pattern from a vertical antenna.*

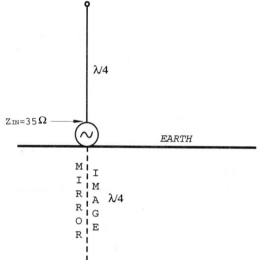

Figure 9-9. *A quarter-wave vertical with earth providing the other quarter wavelength.*

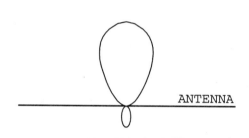

Figure 9-10. *The radiation pattern of the average Yagi antenna.*

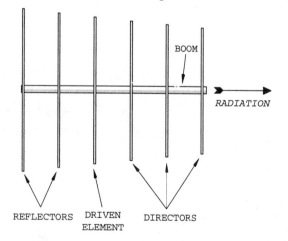

Figure 9-11. *Construction of a popular Yagi antenna.*

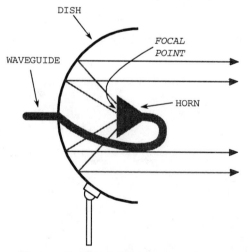

Figure 9-12. *A high-gain microwave dish antenna.*

is fed by the transmission line with the radio frequency energy from the transmitter, while one or more progressively longer *reflectors* reflect back to the driven element an inphase signal that adds to the driven element's radiation. With the addition of one or more progressively shorter *directors* in front of the driven element, the radiation pattern is further enforced, resulting in an extremely directional, high gain antenna. Both the dipole and the Yagi can be vertically polarized if the frequency is high enough to allow the antennas to be small enough to be so physically oriented.

Another classification of the directional antenna is the *driven array*, which uses two or more driven elements fed with a certain phase relative to each other, with each element placed at set distances from one another to create either a bidirectional or unidirectional beam pattern. *Phasing harnesses* or *quarter-wave matching stubs* can be used to ensure the proper phasing to these elements. Common driven array examples would be the bidirectional *collinear array*, the bidirectional *end-fire array*, and the omnidirectional and wide bandwidth *log-periodic array*.

Special antennas are regularly employed in the microwave region. There are two very popular antennas used in the mid- to high-microwave frequency range: the *dish* and the *helical* antenna. The dish antenna (*Figure 9-12*) utilizes a spherical or parabolic shaped focusing bowl with a horn antenna set at the focal point. The dish can be constructed of a solid conductor (usually sheet metal) or wire mesh, with holes 1/8 of a wavelength or less, which act as a solid shield. The horn is attached to a waveguide for transmitting or receiving a signal. The horn itself is merely a flared out portion of the waveguide, similar to a trumpet's bell, which acts as an impedance match between the waveguide and the surrounding space. Gain is dependent on the size of the dish, with the diameter customarily designed to be ten times its wavelength of interest to obtain high gain, low loss, and extreme directivity. The helical antenna (see *Figure 9-7*) is made of conducting wire or tubing wrapped in a helical manner around a nonconducting center strut, typically fed by a short run of coax placed through a ground plane reflector. The output of the antenna is circularly polarized and, as stated above, the receiving antenna must be wound with the same sense as the sending antenna or little or no reception is possible. Gain and beam width of the helical does not measure up favorably with the dish antenna, but the helical is inexpensive and easy to construct for non-demanding communication applications.

9-2. Transmission Lines

Transmission lines, simply, are conductors designed to transfer current from one point to another, without radiating, at a desired impedance. There are two types of RF transmission lines: *balanced*, such as twin-lead; and *unbalanced*, such as coax. In the microwave region *waveguides* are sometimes employed instead of coaxial cables or twin-lead. In addition, impedance matching considerations and matching networks between a transmitter, the transmission line, and its antenna, are vital considerations to lower reflected waves and increase the effective radiated power output.

9-2A. Transmission Line Types

Balanced lines are typically 300Ω twin-lead. (*Figure 9-13*.) Unlike unbalanced line (coax) there is no single conductor at ground potential, and each conductor on balanced line has an equal-in-amplitude but

Figure 9-13. *Construction of twin-lead transmission line.*

Figure 9-14. *Construction of flexible coaxial cable.*

Figure 9-15. *Voltage, current, and impedance on a flat line (no standing waves).*

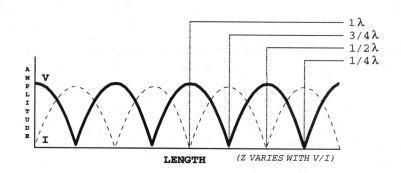

Figure 9-16. *Voltage and current on an open, or non-terminated, transmission line (maximum standing waves).*

Figure 9-17. *A section of microwave transmission line — waveguide.*

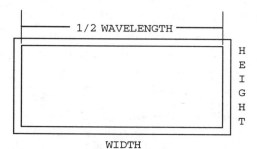

Figure 9-18. *The width of a waveguide sets its working frequency limits.*

opposite-in-phase signal traveling in each line. Twin-lead is generally used as a feed line to a TV or FM consumer receiver or, rarely, as a balanced feed to a dipole transmitting/receiving antenna. It has very low line losses, is capable of supporting high voltages (high VSWR) on the line before arc-over — but is not commonly available in the proper impedance for most transmitters and receivers without matching networks.

Unbalanced line, such as standard coax (*Figure 9-14*), is shielded to prevent receiving or radiating any signal from the line. The outer shield is at ground potential and the inner conductor carries the RF current. Coax can be of the flexible or rigid type. Flexible is by far the most prevalent and comes in two basic surge impedances of 50 and 75Ω, with many other characteristics, such as losses per foot, size, and breakdown voltages of equal importance when choosing coax. These characteristics are shown for each common coax type in *Table 9-1*.

Special coax is available that can be used at frequencies up to 10 GHz.

9-2B. Reflected Waves

As stated, with the transmitter's output impedance matched to the transmission line, and the transmission line matched to the input impedance of the antenna, no standing or reflected waves will be present in the transmission line, and no power will be wasted as heat (except in the pure resistance of the line as copper (I^2R) losses). With all impedances matched, the transmission line appears infinitely long, causing no standing waves to be reflected back into the transmitter finals while transferring maximum power to

COMMON COAXIAL CABLES WITH SPECIFICATIONS AND APPLICATIONS

TYPE	IMPEDANCE (W)	OD (IN)	LOSS AT 100 MHz (dB/100ft)	APPLICATIONS
RG-58	50	0.196	5	CB/HAM — SHORT RUNS, LOW POWER/FREQ
RG-59	75	0.242	3	CB/HAM/TV — MEDIUM POWER/FREQ
RG-8	50	0.405	2	HF/VHF/UHF — LONG RUNS, HIGH POWER
RG-11	75	0.405	2.5	HF/VHF/UHF — LONG RUNS, HIGH POWER
RG-214	50	0.425	1.9	HF/VHF/UHF — LONG RUNS, HIGH POWER

Table 9-1. Common coaxial cables with specifications and applications.

Figure 9-19. *A delta-match for impedance matching between a feedline and its antenna.*

Figure 9-20. *Impedance matching using an adjustable half-wavelength stub.*

Figure 9-21. *A Q-section used for impedance matching.*

Figure 9-22. *Pi-network impedance matching.*

Figure 9-23. *Matching a quarter-wave vertical to its feedline.*

the antenna. This type of line is referred to as *flat line*. (*Figure 9-15*.) When there are high standing waves present on the transmission line, as in *Figure 9-16*, a very high SWR condition exists. This may damage the dielectric and perhaps the finals of the transmitter. (Especially susceptible are transmitters with unprotected transistor finals.)

9-2C. *Waveguides*

A *waveguide* (*Figure 9-17*) is a microwave transmission line. It can be a round or a rectangular hollow metal conduit, usually plated internally with silver, made to conduct microwaves from one point to another with a minimal amount of signal loss. The size of the guide determines its working frequency. (*Figure 9-18*.) Straight 1/4 wavelength probes or loop probes are commonly used to inject or remove the microwave energy from a waveguide.

Waveguides will propagate radiation above their working frequency, but not below their cutoff frequency, basically functioning as a type of high-pass filter.

Wherever feasible, modern microwave designs have eliminated much of the use of waveguides in favor of low-loss semirigid coax cable, with 100% solid-copper shielding, to transmit and receive these high frequency signals.

9-2D. *Antenna Matching Networks*

To ensure the proper impedance match (and thus the proper power transfer) between the feedline and the antenna, or between the transmitter and feedline, some type of matching network may be necessary. The *delta-match* (*Figure 9-19*) technique is applied by tapping, with balanced line, a dipole at a point of matching impedance. Since $Z = V/I$, there is always a point along an antenna that will match the feedline. *Half-wave matching stubs*, which are sections of coax or twin-lead shorted at one end, with the other end connected at the input to the antenna, can be used for this purpose. Both the stub's length, controlled by the *shorting bar*, and the antenna's feed point, controlled by the *slider*, can be made changeable. (*Figure 9-20*). By moving both, a combination can be found that will match most impedances. Quarter-wave sections of coax or twin-lead, called *Q sections*, are also commonly utilized to match small impedance mismatches. (*Figure 9-21*.) By using a section of 50Ω coax between the 37Ω antenna input impedance, the 75Ω coax will be matched to the antenna's feed point. A Q-section value can be calculated by:

$$Z \text{ of } Q \text{ Section} = \sqrt{Z_{LOAD} x Z_{TRANSMISSIONLINE}}$$

The π-network of *Figure 9-22* is frequently employed to tune out all reactances, leaving the proper pure resistance between the feedline and the antenna. It is primarily used in the finals of tube and tunable transistor transmitters. (See *π-Network*). Another method for matching quarter-wave verticals is shown in *Figure 9-23*. The antenna slider moves the feedline's insertion point up and down as the variable C_1 is adjusted, creating a proper impedance match. A *transmatch*, with one type shown in *Figure 9-24*, uses variable capacitors and/or variable inductors to cancel any reactances in the antenna circuit while changing the antenna's resistance, as seen by the *transmission line*, to an appropriate value (usually 50

Figure 9-24. *One form of a transmatch. Employed for impedance matching.*

Figure 9-25. *A balun placed between a transmitter's finals and coaxial transmission line.*

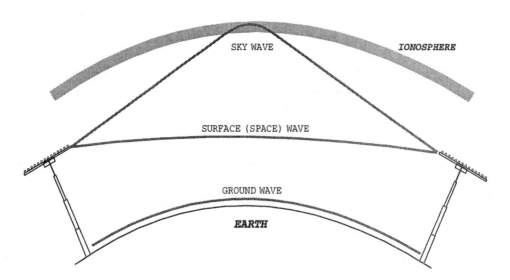

Figure 9-26. *Three modes of RF propagation.*

or 75Ω). If inserted between the transmitter and feedline the transmatch will allow the transmitter to see only its proper characteristic impedance. However, any standing waves on the transmission line itself will not be improved. *Baluns* (*Figure 9-25*) are also commonly utilized for impedance matching. Most are wideband RF untuned transformers adopted to match balanced to unbalanced line, and visa-versa, by isolating the two types of feedlines while matching any impedance mismatches.

9-3. RF Propagation

The method of propagation of radio waves is dependent on the frequency of the carrier and is by three main modes (*Figure 9-26*): *ground waves*, which travel along and through the surface of the earth; *surface waves*, sometimes referred to as *space* or *direct waves*, travel through the atmosphere in a relatively straight line from transmitting antenna to receiving antenna; and the *sky wave*, produced by a radio signal bouncing (refracting and reflecting) off of the ionosphere.

Only very low frequencies are propagated by ground waves (below 1 MHz) to any significant degree, while surface waves are the primary form of propagation for 30 MHz and above, and extend only to slightly more than line-of-sight (*LOS*) distances. These waves propagate about 1/3 farther than LOS due to the earth's atmosphere creating a bending effect of the radio wave, forming a longer *radio horizon* than an optical horizon.

Sky waves are the principal way low and high powered simplex long range RF communications can take place. The frequencies utilized are essentially under 30 MHz, and the *optimum usable frequency* (OUF) varies tremendously with the time of day, sunspot activity and their eleven year cycle, and the angle-of-radiation from the transmitting antenna. The optimum usable frequency is the frequency at which the most reliable communications can take place via the sky wave. It is always between the *lowest usable frequency* (below the LUF the ionosphere absorbs the signal) and the *maximum usable frequency* (above the MUF the signal passes through the ionosphere and out into space). *Multiple-hop transmissions* caused by the transmitted RF signal bouncing more than once off of the ionosphere can occur, extending the communications range of a radio signal to thousands of miles—with relatively little output power.

Troposcatter is a form of dependable long range communications (about 450 miles) using high-powered UHF transmitters with high gain antennas. These transmitters send a powerful signal into the troposphere (a section of the atmosphere about six to ten miles above sea level), where the signal scatters. A minute portion of that scattered signal is picked up by a very high gain receiver and amplified.

Due to the high cost of the equipment, the military and large industry are the primary users of this communications method.

CHAPTER 10
Troubleshooting

Troubleshooting is a vital part of the technician and field engineer's job. Unless the equipment and circuits that are being repaired are well understood, troubleshooting down to the component level will be a very time consuming job. While the all too common underskilled board-swapper has taken over the lowly tasks of the old vacuum tube-swapper, a technician that can actually troubleshoot down to the component level is a very rare and valued breed. It demands little skill to find a board that displays a red light, indicating a bad board, in a piece of complex equipment. On the other hand, it requires exceptional skill and knowledge to locate the actual *component* that caused that fault. This skill can be obtained by studying as well as, of course, practical experience.

10-1. General

Keep in mind that *all* power amplifiers, whether RF or audio, must be terminated by their output imped-ance (Z_{OUT}), or they can be destroyed when activated. This is accomplished by the use of a resistor, dummy load, or other appropriate resistive terminator. Only then can the amplifier or transmitter be operated and the relevant tests conducted.

Also, many new technicians feel that a sudden fault in a transmitter or receiver, such as low receive or transmit output power, is caused by misalignment of its stages. Seldom is this the case. Lack of alignment is gradual and usually caused by aging components. However, any attempted repair or re-alignment by an incompetent or inexperienced technician can cause a receiver or transmitter to need retuning.

10-1A. Safety

Safety is always of paramount concern, since one fatal mistake will be your last.

Basic safety rules:

1. When working on high current and/or high voltage circuits, place one hand in pocket when probing live circuits.
2. Use an isolation transformer when working on transformerless equipment.
3. Discharge all high current/voltage capacitors.
4. Try never to work alone.
5. Never place yourself at ground potential when working on live equipment.

10-1B. *Basic Troubleshooting Steps*

The basic steps in electronic troubleshooting are: (A) *define the symptoms* (low power, no power, off-frequency, distortion, etc.); (B) *restrict the fault to a section* (the IF strip, the frequency converter stage, the audio amplifiers, the power supply, etc.); (C) *track down the fault to a single bad stage* (the driver, the first IF stage, the LO, the final power amplifier, etc.); (D) *trace the fault to a bad part* (capacitor, transistor, relay, solder joint, etc.).

First, obtain the service literature and, if you are very lucky, they will contain two or three of the following: a *functional block diagram* (each transistor in the device compromising a separate block); a *servicing block diagram* (each transistor comprising a separate block as well as pictorial waveforms and voltages at test points); a *schematic diagram*; and a *parts layout diagram* (which shows the actual physical layout of the parts).

Step A. *Define the symptoms.* First make sure the unit-under-test is plugged in, *that all operator controls are in their proper positions*, and that there are no exterior circuit breakers or fuses blown. If fuses are blown, then replace fuses with proper amperage rating as well as the proper voltage rating (which is the maximum rating, when the fuse is in the open condition, before arcing between its contacts occur). Confirm the fuse does not blow again under full operating conditions. If it does, then open the unit's case for further troubleshooting.

Operate the device and observe incorrect operation or, employing test equipment, search for improper frequency response, distortion, power output, etc.: Is the transmitter section of the transceiver bad? Or is it the receiver section?

Step B. *Restrict the fault to a section.* First, crack the case and *look* for burnt components, broken traces, bad solder joints, smoke, etc. *Smell* for burning resistors, transistors, and overheated transformers. *Touch* (carefully), checking for excessively hot components. *Listen* for voltage arcing or sizzling of burning parts. This preliminary inspection is extremely important! Also, since modern transceivers usually employ some sort of *circuit sharing*, keep in mind that a fault in a single circuit can take out both the transmitter and receiver sections. If the fault is not obvious, then bracketing must be used along with *signal injection* or *signal tracing*. *Bracketing* refers to taking an entire section that has a good input, but a bad output, and continuing to move your probe from section to section until the problem is confined, within these imaginary brackets, down to the bad area. For example, a receiver could be bracketed until a lack of receiver output is discovered to be located in the IF strip of a receiver, which may contain a chain of three or four amplifiers and numerous filters. (*Figure 10-1.*)

Step C. *Track down the fault to a bad stage.* Use the functional block diagrams to isolate the problem to an individual circuit by narrowing down the probable bad stage by continued bracketing from stage to stage, instead of from section to section — such as from the speaker to the audio power unit of *Figure 10-2.* In addition, use all available schematics as needed and check voltage levels and waveforms at the appropriate test points. For example, is the voltage amplifier, driver, or power amplifier in a non-functioning audio amplifier chain bad?

Step D. *Trace the fault down to a bad part*. Continue to utilize sight, touch, smell and hearing to check for burnt, overheated, cracked or arcing parts. Employ a scope to observe waveforms and a VOM to check voltages and continuity while referring to the schematic diagram. Check any bias voltages and calculate currents in the bad circuit. If the schematic is not marked with the proper voltages, use your electronics knowledge to estimate the proper amplitudes. Do these measured amplitudes vary from your own calculations or other similar circuits? If an individual component is now suspected, make an in-circuit test by checking for the component being possibly shorted or opened. These simple tests assume that the technician has reasonably ruled out other defective components shunting current to ground, or delivering high current levels to the suspected component, or from being open in the suspected component's current path. To find a short in a component take a voltage reading across the component — if the component is shorted, then there will be little or no voltage dropped (if a voltage drop is expected). Check for an open by observing a large voltage drop across the component (if little or no drop is expected). If no parallel paths are present to fool an ohmmeter, then take a simple resistance reading on unpowered equipment across the component. Then, if the component is still suspected, test the part out of circuit. If defective, replace the component with one that is known to be good. Power up unit to check for continued overheating, improper operation, poor waveforms, bad voltages, etc.

The *servicing block diagram* of *Figure 10-3*, or the *schematic* of *Figure 10-4*, with marked AC waveforms and DC bias voltages, if available for the unit-under-test, make troubleshooting much easier. It is impossible to effectively troubleshoot a modern receiver, transmitter or transceiver without at least a simple schematic or block diagram — unless the problem is visually apparent or unbelievably obvious. Keep in mind that all indicated AC and DC voltages in such diagrams are usually referenced to ground, and not across a semiconductor junction or component. As well, all indicated voltages are only approximate, since most commercial quality components have rather loose tolerances, not to mention the variable characteristics of semiconductors, even when identical to each other.

Further tests and procedures follow to help in troubleshooting and repairing modern RF equipment.

10-1C. Signal Tracing

A common technique in troubleshooting receivers or amplifiers is *signal tracing*. (*Figure 10-5*.) To perform signal tracing on a receiver, warm up the receiver, the frequency generator, oscilloscope or frequency analyzer, for twenty minutes. Confirm that the receiver is tuned to the generator's output frequency. Place the antenna or probe of the modulated RF generator near the receiver's antenna or, if de-tuning of the front-end does not result, attach the probe directly to the receiver's antenna input. Start with a generator output of about 100 μV and decrease, if possible, to avoid amplitude distortion. Using an oscilloscope or spectrum analyzer with probe, check each stage's output for the proper level and linearity. No signal, or a distorted signal, at the output means the stage, or its AGC voltage, is defective.

To *signal trace* a low powered transmitter, turn on the carrier or, with SSB, modulate with a single or dual tone, and trace from the audio and master oscillator stages to the output, being careful of the final power amplifier stage.

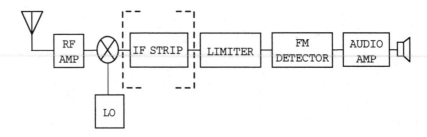

Figure 10-1. Troubleshooting by bracketing down to a section.

Figure 10-2. The process of bracketing. First check for a bad speaker (a), then backup to the power amplifiers (b).

Figure 10-3. A servicing block diagram showing proper waveforms and amplitudes.

10-1D. Signal Injection

A similar technique to the above is called *signal injection.* (*Figure 10-6.*) To accomplish signal injection, warm up test equipment and receiver under test. Inject the proper frequency, at the proper *amplitude* (to eliminate amplitude distortion), beginning at the speaker and heading back toward the front end. Listen for a speaker output. The stage that is defective will not send a signal to the speaker; or will have a low output; or will introduce distortion. You *must* change the signal generator's output frequency for each stage being injected. (For instance, an AM IF stage would need a signal injected at 455 kHz, while the audio stage would need around 1 kHz, etc.)

When injecting the test signal into an AM or FM receiver, the application of *standard test modulation* values are employed. For AM, a 1 kHz tone of sufficient amplitude to create 100% modulation is used; while for FM, a 1 kHz tone of sufficient amplitude to create a 3 kHz deviation has been adopted. SSB receivers use a *two-tone* modulation, with each tone being of a matched amplitude. Each tone can be anywhere between 300 Hz and 3 kHz, with a separation of at least 1 kHz between tones and not harmonically related to the other.

10-1E. Lead Dress

Lead dress (*Figure 10-7*), which is the physical layout of connecting wires, becomes a major consideration in circuits used in high frequency applications. This is due to the natural capacitance and inductance of the wires. After repair or alignment, this lead dress must be maintained or unpredictable circuit operation, or actual de-tuning of tuned circuits, can result. As well, lead lengths of components must be kept short for maximum power dissipation ratings.

10-1F. Heat Sinks

Never apply power to circuits under test without the component's proper heat sinks in place — or without the component or IC bolted down to the chassis that is acting as a heat sink. Rapid thermal damage to the component can result. Furthermore, capacitance is added to or subtracted from a circuit if an RF transistor is separated from its heat sink. Also, use only the same thickness insulators between an RF transistor and its heat sink as in the original equipment, or circuit instability or frequency shift might result. This is caused by a change in stray capacitance.

10-2. Intermittents

One of the most difficult and frustrating troubleshooting jobs can be the location and identification of an intermittent problem. A simple and very effective technique to check for intermittent operation of a device caused by bad components, cracked PCBs, or bad solder joints, is to thermally stress the suspected circuit areas. While the unit under repair is turned on, chill the general area with freeze spray while checking for proper circuit operation or voltages. A stubborn intermittent may also have to be further stressed by the application of heat immediately after the circuit is refrigerated. Employ a stan-

Figure 10-4. *A schematic with DC voltages and AC waveforms.*

Figure 10-5. *The signal tracing method of troubleshooting.*

1. AUDIO FREQUENCY

2. MODULATED IF

3. LOCAL OSC FREQ

4. MODULATED RF

Figure 10-6. *The signal injection method of troubleshooting.*

dard heat gun, switched to low heat, while utilizing a heat concentrator (reducing or constrictor) nozzle for local application. If the unit was already functioning properly, improper operation should now become evident. If the unit was faulty when received, temporary proper operation can result. Either way, localized thermal stresses placed at suspected areas allows troubleshooting of intermittents to be accomplished quickly and simply, down to the component level.

Intermittent problems in circuits are commonly caused by capacitors, especially electrolytics, or broken traces, or cold solder joints. (Bad solder joints can create a high resistance in the effected joint, producing unusual and unexpected circuit operation.)

10-3. Parts Failure and Their Indications

ICs, since they are almost 50 times more reliable than separate discrete circuits, are very dependable — and rarely fail. Suspect them, while troubleshooting, as being internally damaged, dead last. Decoupling and coupling capacitors, on the other hand, fail often. Decoupling capacitors in an amplifier stage (which prevent RF or AF from feeding into the power supply) if open, can cause self-oscillation in that or other stages. If shorted, they can short out the power supply. A shorted coupling capacitor can affect the bias on the next amplifier stage. As a general rule, a shorted capacitor will effect DC and can affect AC; while an open capacitor will not affect DC voltages, but will affect AC voltages. Bridging a suspected open coupling or bypass capacitor with another capacitor may find the problem. Also, an excessively warm capacitor can indicate that there is DC current flowing through it, creating power dissipation in the leaky, or partially shorted, dielectric. $(P = I^2R.)$

When dealing with questionably bad resistors, bear in mind that when a resistor becomes defective it will *usually* open, or partially open, creating a high value of resistance. They will rarely internally short, or become lower in resistance values.

If a defective transistor is found, then examine other components that could have contributed to the failure, as transistors rarely go bad unless a defective passive component creates the failure — or, possibly, it can be due to poor circuit design, allowing too much current to flow through the semiconductor. A bad component can commonly force the transistor to conduct excessively high currents, damaging it. Also, when an amplifier is not functioning properly, the problem is usually in the DC bias voltages — check biasing components for any changes in value. As well, transistor failure can be caused by an overly high drive signal. Or a preceding stage may become defective, adversely affecting the stage under test. For example, in an RC or LC coupled two-stage amplifier, the coupling capacitor between stages may short, driving the second amplifier stage into saturation by the direct application of the Vcc of the first stage to the base of the second stage, creating excessive bias. A shorted emitter bypass capacitor or, much more rarely, the emitter resistor itself becoming internally shorted, can create excessive current (depending on amplifier design) through the decrease in negative bias and the resulting thermal instability, slowly causing temperature related damage, lowering beta and increasing collector leakage current. A high Vcc, ordinarily caused by defective power supply regulation, can destroy a transistor, as can voltage transients (especially during thunderstorms). Some of these component faults are cumulative and slowly destroy the transistor over time, while gradually creating output distortion due to the shift in the amplifier's Q-point bias. Other faults may immediately destroy the semiconductor within microseconds.

10-4. Linear IC Troubleshooting

Individual components in an IC cannot be examined, and (when confronted with many IC leads) troubleshooting of integrated circuits can sometimes be quite daunting to the inexperienced technician. So much so, in fact, that many technicians simply replace the IC before properly checking out other options. ICs have become very reliable, and are extremely dependable — far more so than discrete circuits. This means that the IC itself is seldom defective.

First, confirm that the suitable ripple and glitch free positive and/or negative supply voltage(s) are reaching the Vcc pin or pins, as well as the proper ground to the ground pin. (Is there a zero resistance path to the grounding point?) Verify that the proper input signal is being inputted into the proper pin. Also confirm that the output signal is present at its appropriate pin. If the input is good, but there is no output, or an improper output, check the following for defects: support components, such as open or shorted coupling or bypass capacitors, defective or damaged frequency determining crystals, bad voltage reference Zeners, etc., which can be pulling down the IC's output, or giving the incorrect output. In addition, a support chip may be feeding improper voltage levels or waveforms to the chip; or the chip's secondary output pins (such as outputs from the chip's internal LO and mixer) may be shorted, possibly affecting the chip's main output.

When all support components have been tested and found not to be at fault, then remove the IC and replace with a known good chip. Test for proper function.

10-5. Amplifier Self-Oscillation Troubleshooting

Any amplifier that breaks out into oscillations creates undesirable results, such as lowered output power, overheated parts, and distortion. Two amplifier self-oscillations are encountered: *feedback* and *parasitic oscillations*. The most distinct difference between feedback and parasitic oscillations is that parasitics can occur at almost any frequency, while feedback oscillations are strictly at the amplifier's signal frequency.

10-5A. Feedback Oscillations

An RF amplifier may begin oscillating at its operating frequency due to its internal capacitance feeding an inphase feedback signal back to its input at a certain frequency. Also, undesired external coupling among the input and output components and wiring can cause oscillations, usually demanding shielding between the input and output circuits to stabilize the stage. An unneutralized RF stage can act as a tuned-collector tuned-base oscillator due to these input and output tuned circuits. Increasing the frequencies that the amplifier must boost is more likely to cause regenerative feedback, producing oscillations, spurious outputs, distortion, overheating of parts — and an increase in plate or collector current.

If a neutralized amplifier stage begins oscillating, check that the neutralizing capacitor or inductor is not open, or that a new feedback path has not been created by conductive dust, fragments or grime (such as soda pop, metal shavings, or a solder blob). If these are not present, then adjust the neutralizing capacitor or inductor as indicated below. Also, a stage will need readjustment of its neutralization if a new tube or transistor has been inserted, since the active device will almost certainly have a slightly different feedback capacitance value.

One type of neutralization procedure employs a tunable inductor, L_n. (*Figure 10-8*.) The inductor is placed between the collector and the base, with a high value capacitor in series with the inductor to avoid a DC short. The inductor is now in parallel with the collector-to-base capacitance, effectively forming a tank circuit. The inductor is tuned until resonance occurs, causing the tank to be a very high resistance at the undesired feedback frequency, eliminating the inphase feedback voltage.

Neutralization is also accomplished by adding a capacitor (C_n, *Figure 10-9*) from the transistor's output back to its input. This supplies a bucking out-of-phase feedback voltage to cancel the inphase feedback voltage creating the offending oscillations or amplifier instability. Tuning the capacitor until the positive feedback is canceled is described below.

At the unneutralized and *unpowered* amplifier stage, inject the proper amplitude and frequency signal into its input while monitoring the collector or plate tank circuit for this signal (a spectrum analyzer or an oscilloscope, equipped with a loosely coupled *sniffer* probe, can be employed to detect the presence or lack of oscillations) — if present, then adjust the neutralizing capacitor or inductor until the stage stops passing the signal from its input to its output, indicating full neutralization. As well, excess collector or plate current and spurious emissions and distortion will all decrease when neutralization is reached. The neutralizing capacitor is generally adjusted until it is of the same value as the collector-to-base capacitance. Another common technique to perform neutralization is to inject a drive signal at the amplifier's design frequency into the *output* of the amplifier, while some type of RF test unit, such as a spectrum analyzer (equipped with a special high input impedance probe matched to the low input impedance of the analyzer's front end, or use the common sniffer probe) or an oscilloscope, is attached to the *input* of the unpowered amplifier. C_n or L_n is then adjusted for a null indication on the display, indicating a cancellation of the positive feedback from its output into its input.

High powered tube amplifiers may use a grid-loading method to neutralize a stage, which effectively lowers stage gain but avoids the need to employ a neutralizing capacitor or inductor. This is commonly accomplished by a grid-to-cathode resistor, or a low tap on the grid tank coil.

10-5B. *Parasitic Oscillations*

Parasitic oscillations are caused by stray capacitances and inductances, created by coupling capacitors, coils, leads, or active devices, forming tuned circuits within an amplifier, making the amplifier act as an oscillator. These oscillations are at a higher or lower frequency than the amplifier's design frequency. In fact, the parasitic oscillations can reach into the UHF spectrum, even if the transmitter's operating frequency is 30 MHz. The frequency of the oscillations depend on the value of the stray capacitances and inductances.

Low frequency parasitics are usually aggravated by a transistor's increasing gain as the frequency is lowered. This produces low frequency instability in an amplifier, since any low frequencies naturally have a higher gain than higher frequencies in any active device.

Parasitic oscillations cannot be treated by a neutralizing capacitor or inductor. The cure is to lower the Q of these undesired parasitic circuits with added low-value resistances in series with the inductors; or use ferrite beads in the grid or base to suppress any VHF or UHF parasitic oscillations; or utilize a small value choke in series with the grid or plate circuit of a vacuum tube RF power amplifier (to apply any reactance, if not so designed by the OEM, to the output of a (typically) low output impedance RF power *transistor*'s collector can have disruptive consequences, and is not recommended); or extend the plate lead of the power tube for added inductance. Low frequency parasitics can be suppressed by a low-pass filter added between the collector and the base of the offending amplifier, which feeds a progressively low frequency degenerative out-of-phase signal back to the input, thus decreasing low frequency gain. Any component so used to eliminate parasitics is referred to as a *parasitic suppressor*.

A word of caution*: It is strongly recommended that a technician should never add additional components, or modify a transmitter or receiver, without the expressed permission of the supervising RF engineer.*

Use a spectrum analyzer with a sniffer probe to detect parasitics. Any unexplained frequencies can indicate parasitics. Also, when checking AM trapezoidal patterns on an oscilloscope, any unexplained nonlinearities in the patterns can mean parasitics.

In a well-designed transmitter little or no parasitics are present, but after repair (such as tube replacement) or modification, they can arise. Simply confirming that lead dress and length are as designed and, if not, reworking to specs, will cure parasitics 80% of the time.

A transmitter with parasitics will waste power and can create internal arcing, overheating, blown fuses and relays, destruction of capacitors, tubes (caused by increased plate and grid currents) and transistors, and can produce erratic transmitter tuning and the creation of interfering (spurious) signals.

10-6. Receiver Troubleshooting

Troubleshooting AM and FM receivers, as well as their associated AGC and specialized circuits (such as squelch, frequency synthesizers, and battery saver circuits), can appear to be quite daunting. However, if a certain logical procedure is followed, most problems can be determined and repaired quickly.

The reason behind total failure of a receiver to perform its function is usually much easier to discover than the sometimes subtle loss of selectivity, sensitivity, dynamic range, or S/N.

10-6A. Receiver Interference, Distortion and Noise Troubleshooting

Interference and noise in a receiver can be caused by: *inadequate selectivity*, since wide bandwidth equals more natural noise pickup and increases the chance of adjacent channel man-made interference; by *regeneration* (positive feedback) in an audio amplifier, causing (low frequency) noise-producing oscillations (referred to as *motorboating*, which is a "put-put" sound that can also occur as a consequence of a low battery in some portable radios); by *noise* on the power line feeding the receiver, producing hash type noise interference (cured by the use of a low-pass power line filter); by *RF transmission line noise* pickup induced by damaged or poorly shielded transmission lines (utilize only high quality shielded and undamaged lines); by a poor *ground*, causing a high impedance path to ground for stray RF and noise, and intermittent ground, creating high noise levels, or *ground loops*, producing low frequency noise pickup through induction caused by power transformers, switching power supplies, brush type motors, etc. (employ proper size ground straps of short length to an adequate ground, well bonded to the grounding rods, and remove all ground loops and bond to a single ground point); by an improperly shielded or leaky *receiver case*, allowing interfering signals into the internal high gain amplifiers stages, causing intermodulation distortion (which then produces many noise generating and non-harmonically related new frequencies) and desensitization (creating extra noise due to the need to increase gain in RF and audio amplifiers in an attempt to overcome the desensitization); by a low frequency hum caused by faulty filtering or regulation of the power supply, or poor shielding of the power supply's power transformer, that feeds the effected receiver; by bad (cold) *solder joints*, creating a high resistance connection where noise is dropped across; by unbalanced push-pull amplifiers in the audio section causing noise; by intermittently bad *connections*, such as faulty cable or socket connections, which can produce popping or scratching noises (physically tap areas that are suspect for any increase in noise output); by bad or failing *components*, which can create noise that is usually sensitive to physical impact or temperature variations (tap suspected components, or freeze components with freeze spray — if noise level changes, the component just tapped or frozen is the culprit); by a *receiver* with squelch, since a bad squelch can cause a lack of muting when no signal is being received — thus amplifying any noise present.

If there is distortion in a receiver when listening, first check the speaker — the paper cone could be torn. Check for crossover distortion (*Figure 10-10*) in any push-pull amplifiers; crossover distortion is produced by a lack of true class B operation caused by improper bias. Any class A amplifier that reaches either saturation or cutoff produces *amplitude* distortion, which can be caused by a change in its bias, either through DC bias or AGC component changes, or power supply variations. If amplitude distortion occurs only during strong signals, AGC is bad, or an output transformer is becoming saturated and is unable to pass any increase in signal level, creating the distortion. Bad LC or RC components in a filter network can create *frequency* distortion, which amplifies some frequencies more than others outside of its design parameters. As well, frequency distortion can be produced in an AM or SSB receiver if the LO drifts off frequency, causing *sideband cutting* to either the lower or upper sideband. A damaged transistor may likewise introduce distortion, as can positive feedback through the interelectrode capacitance between its collector and base. (See *Self-Oscillations*.)

10-6B. AM Receiver Troubleshooting

Turn on the radio at high volume. If there is only white noise, which is circuit generated (mainly by the RF sections), even when tuned through its entire band, then the audio and RF amplifiers are probably good. Check the local oscillator or PLL for proper frequency and amplitude output, then mixer, IF, and detector. If a receiver is not receiving at the indicated dial frequency after signal injection, then the LO or PLL is probably off frequency. If the LO is oscillating, then examine the antenna tuning capacitor for a warped or shorted rotor or stator, or if the RF antenna coil is open. A LO may also become loaded down, causing a frequency shift, if the next stage (usually a mixer), becomes defective. Also, a low supply voltage caused by a bad regulator can cause the LO to shift in frequency or become unstable.

If the receiver is completely dead, confirm that the proper voltage is reaching the PS's power transformer. Check fuse and voltage outputs of the power supply. If the power supply output is low, the power supply leads should be disconnected from the receiver and tested, as a receiver fault could be pulling down the supply voltage, or current limiting could be kicking in too early in the PS regulator circuit, furnishing low current levels to the receiver. If the output voltage or current level of PS is still bad, troubleshoot the power supply. If all PS voltages and currents are acceptable, then increase the receiver's volume and use the half-split signal injection method by injecting an audio signal at the beginning of the audio stages (customarily at the volume control). If no signal is heard, then confirm that the squelch circuit is set and operating properly, and troubleshoot down the audio stages to the speaker. If a signal is heard at the speaker after injection at the volume control, then inject a modulated RF signal of the correct (dial) frequency into the antenna. If there is now no speaker output, then move the injection point, changing frequency as required, toward the audio stages, until a tone is heard. The stage before that point is the defective stage. When the faulty stage is localized, then be sure to troubleshoot to the component level.

If the output is distorted, insert an audio tone at the first audio stage, then check for distortion at the appropriately terminated output of the last audio stage with an oscilloscope. This distortion will usually be in the form of clipping of the peaks of the audio waveform. If these are present, then the bias levels of one of the audio stages may be faulty. If clipping of the signal occurs before the detector, the AGC could be defective, or there may be very bad IF amplifier DC bias levels. Distortion in amplifiers can also be caused by a *prior* stage feeding an excessively large input signal into the amplifier.

If a receiver stage has a low output, the bypass capacitor across the amplifier's emitter resistor may be open; or a coupling capacitor to a next stage or circuit may be open; or AGC may be defective; or the active device of the amplifier could be faulty. Possibly amplifier self-oscillations are occurring, creating a strong AGC reaction to what may only be a weak input signal. Finally, check the alignment of the receiver.

Almost all receivers have AGC circuits that limit gain — so AGC will have to be disabled by cutting the AGC line, or by injecting a reverse "bucking" voltage into the AGC line, or by turning off the AGC with a panel switch (when so outfitted), if the AGC and the amplifiers that the AGC controls need to be isolated from each other to segregate the bad circuit. However, if the DC bias to the amplifiers is being completely or partially supplied by the AGC at all times as a normal condition of circuit operation, then clamp the AGC line to an appropriate DC voltage and continue troubleshooting. With any receivers having poor sensitivity, check that all AGC voltages are correct. Defective AGC may also completely or partially shut down one or more good RF or IF stages. Also keep in mind that it is sometimes difficult to work with receivers equipped with AGC, since AGC is a loop — a bad IF signal output causes incorrect AGC voltages; while a faulty AGC can cause incorrect IF output.

Always replace all removed shielding, as the shielding of stages or components prevents or reduces interaction between stages that can cause undesired oscillations, degeneration, or IMDs. Coils, tubes, and capacitors, as well as interstage leads, are susceptible to electromagnetic-field coupling, causing interference. Some old, poorly designed, or improperly repaired equipment may need extra or supplementary shielding to attenuate this coupling.

SSB is similar to standard AM except for the need to confirm that the BFO is on frequency and outputting a clean and sufficiently high amplitude signal into the detector. If not, then noisy, off-pitch, or no audio output at all can result.

10-6C. FM Receiver Troubleshooting

See above, except for the following differences. Some FM receivers do not employ AGC in the IF stages, but only in the RF section. Also, do not be mislead into thinking that the last IF stage of an FM IF section has defective DC bias, as this stage is sometimes overdriven to act as a limiter for the detector. Furthermore, if an FM receiver drifts, then the AFC (or PLL) section of the local oscillator stage is faulty, or the PS regulator is failing, causing frequency instability. In addition, handhelds with *private line* must be unsquelched by the proper tone injected into the front end, or no output into the speaker is possible.

10-7. Transmitter Troubleshooting

Transmitter troubleshooting requires much more caution to undertake than working on receivers. Transmitters can generate very high voltages and currents, which can prove fatal to whomever comes in contact. In fact, some high powered transmitters will not allow the transmitter to be internally inspected or repaired with the power on; all access doors may have safety interlocks that shut power down as soon as a panel is opened. As well, the technician must deal with a device that radiates a signal that can, if improperly repaired or aligned, interfere with other RF communications — much to the intense displeasure of your employer and the FCC.

With the appropriate and cautious use of oscilloscopes, DMMs, spectrum analyzers, and SWR/wattmeters, most transmitter troubles can be quickly isolated.

10-7A. Common Transmitter Problems and Troubleshooting Techniques

As mentioned above, transmitters, due to their inherently higher voltages and currents present during operation than those existing in receivers, can be dangerous to troubleshoot in all but low powered

ORIGINAL LEAD DRESS

RIGHT   WRONG

Figure 10-7. *Three PCBs showing right and wrong lead dress.*

Figure 10-8. *Inductor neutralization circuit of Ln and Cb.*

Figure 10-9. *Neutralization using an adjustable feedback capacitor Cn.*

CROSSOVER DISTORTION

Figure 10-10. *Crossover distortion as viewed on an oscilloscope.*

FUNDAMENTAL

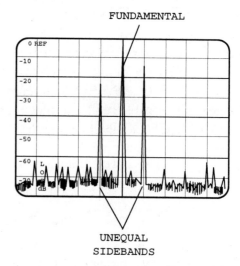

UNEQUAL SIDEBANDS

Figure 10-11. *Incidental FM as viewed in the frequency-domain.*

(QRP) transceivers. However, the larger transmitters generally have external panel voltage and current meters that can help the technician to isolate most problems down to a stage, while some even have switchable (into a desired circuit) oscilloscopes, signal monitors, and SWR/wattmeters. This allows the transmitter to be troubleshot while under power. In smaller transmitters, and almost all transceivers, it is appropriate, if one follows proper safety precautions, to internally signal trace down to the bad stage while under power.

Whatever the transmitter, first check for a proper signal output by monitoring the frequency accuracy, signal cleanliness (harmonics, IMDs, parasitics), RF power level, and modulation percentage. Examine for the correct power supply voltage and lack of AC ripple to each stage, as well as any low or distorted levels in the audio amplification stages. If a transceiver does not transmit or receive, but there is some white noise during receive, check for a bad PLL. (The frequency readout on a transmitter or receiver may not indicate an error with the PLL.) If a high SWR is shown on the SWR bridge, look for a mistuned transmitter final; a broken or mistuned matching network or antenna; a damaged RF cable; or a corroded RF connector. Even if the SWR does not seem excessively poor, it can appear to be better than it is if the transmission line is lossy. Since the SWR is commonly sampled at the transmitter output and the transmitter power must travel down the lossy line and then return (if not properly terminated), then the reflected power, and the forward power through the line, is consumed as heat. This makes the reflected power lower when it reaches the SWR bridge at the transmitter's output, and reduces the ERP (*effective radiated power*) below what it should be. Also, if one RF amplifier stage in a transmitter is slightly detuned, then phase modulation will occur. This is due to the detuned amplifier tank presenting either a capacitive or an inductive reactance to the signal, creating phase modulation, causing possible distortion at the receiver. As well, general rapid carrier frequency instability can be produced by poor power supply regulation, creating a master oscillator that instantly alters frequency due to a varying supply of power (*dynamic instability*), sometimes only during modulation — or it can be created by actual mechanical vibrations causing momentary changes in component values. This is referred to as *frequency shift* and can be detected by attaching a frequency counter to the output of the unmodulated transmitter and checking for oscillator instability by vibrating the offending transmitter or by stressing the power output of the common supply. FMing (also called *incidental FM* or *residual FM*) is produced by another source (such as noise or AM) modulating the carrier, creating a generally rapid and fleeting, but sometimes continuous, unwanted frequency modulation. If riding on a desired AM signal, incidental FM may show up on a spectrum analyzer display as one sideband being of a higher amplitude than the other (*Figure 10-11*), or as a slight horizontal jitter of the carrier on an oscilloscope display. (*Figure 10-12.*) Excessive (out-of-tolerance) *frequency drift* is a gradual change in frequency over time, caused by defective, or badly designed, oscillators or frequency control circuits (or a blocked or damaged cooling fan causing heat buildup, changing the internal capacitance of the transistors and the values of the LC components).

If the transistor finals of a transmitter are burnt out, first suspect the cause as high SWR. If tube driven, then suspect worn-out tubes.

When RF transistors, tubes, components, or leads are replaced during service, then a retune may be necessary due to the probable changes in the reactances of the sensitive RF circuit.

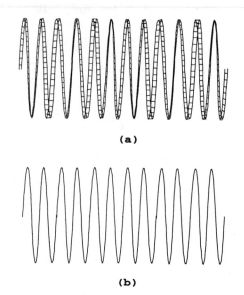

(a)

(b)

Figure 10-12. *A waveform on an oscilloscope with incidental FM (a), and without (b).*

LOOP CURRENTS

UNIT UNDER TEST

Figure 10-13. *Severe ground loop formation while testing.*

Vcc

R_{TEST}
$10K\Omega$ R1 R_C

VOM

SHORTING
WIRE

R2 R_E C1

Figure 10-14. *In-circuit transistor test.*

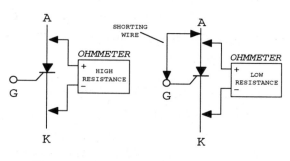

A   A
        SHORTING
        WIRE
              OHMMETER   OHMMETER
              +   +
              HIGH   LOW
              RESISTANCE   RESISTANCE
G   −   G   −

K   K

Figure 10-15. *Testing procedure for an out-of-circuit SCR.*

10-7B. *Transmitter Tube Final Troubleshooting*

In a defective tube circuit the same troubleshooting techniques apply as above except, in view of the high powers always involved with a modern tube transmitter, also check for arced tube pins and sockets (causing a constant or intermittent open, or a high resistance contact), broken tube pins and, of course, bad capacitors, burnt resistors, and bad tubes (failed vacuum tubes can account for up to 70% of tube circuit failures). If the tube's heater is not glowing or, in metal envelope tubes, the envelope is not extremely warm (**Caution**: *Many tubes become extremely hot*) then the heater element, or its power supply circuit, is defective. Remove the tube and test the filament with an ohmmeter for continuity. While powered, check for a leaky tube, frequently evidenced by a green, blue or purple haze inside the tube (a "soft" tube) causing excessive plate current and erratic amplifier operation (a dim dark purple glow is generally normal). Check for the appropriate tube amplifier bias. If the bias goes out in some old unprotected tube circuits, excessive currents will flow and the tube may be destroyed. Also, cooling fans are needed in most power tube circuits due to the extreme heat generated under normal operation. If these fans cease to work or become blocked, the tube will be damaged or destroyed. Moreover, well-used tubes will naturally decrease their current output. This is caused by the cathode's coating flaking-off or becoming pitted (*cathode depletion*) with use or excessive current flow. Further, confirm that the electrodes inside the tube have not become shorted. If the electrodes are not visible, employ an ohmmeter to verify a lack of continuity from electrode to electrode. (Cathode-to-suppressor grid and beam electrode-to-cathode shorts are normal.)

10-7C. *Transmitter Harmonics*

If excessive harmonics are present in the output of a transmitter, then confirm proper operation of the transmitter. Establish that all stages are properly biased and tuned and that none are being overdriven (which causes flattopping and thus harmonic generation), and that all low-pass filters and/or wavetraps are set and in operation. Make sure that suitable grounding of all appropriate devices (such as the transmitter, test equipment, and external filters) is in place. Confirm that any Faraday screens, which are sometimes installed between the final RF power amplifier transformer's primary and secondary to lessen capacitive coupling of harmonics to the antenna, are also grounded. If a transmatch is in use check for proper tuning (to attenuate any frequencies not of the fundamental).

10-8. Common Tests

The following are important tests that allow the technician to confirm proper operation of RF and AF equipment and their associated components, and as an aid in troubleshooting and alignment.

Caution: *Never hook up test or other equipment*, as shown in *Figure 10-13*, since ground loops are formed. A ground loop causes low frequency magnetically induced interference to be picked up by this closed loop. The main culprit is commonly a 60 Hz power transformer transferring its hum to the equipment. Minimize loops by grounding to a single point, thus decreasing noise driven induced voltages by minimizing the loop area.

Warning: *Never use a scope to measure voltages across a component*, or a short to ground will result. The scope probe's ground should always be connected to the unit-under-test's common ground only. When the voltage drop across a component must be measured, then employ a DMM or VOM instead.

10-8A. *Transistor In-Circuit Test*

A simple in-circuit good/bad test will confirm the basic operation of a transistor. (*Figure 10-14.*) Short the base to emitter — the transistor should turn off. (Voltage across the emitter-collector increases.) If it does not, then the transistor is probably bad. If the transistor does turn off, then parallel a resistor across R_1 (which decreases R_1's value, which increases forward bias) — the current through the transistor should increase. (Voltage across the emitter-collector decreases.) If the current does increase, then the transistor is good.

Table 10-1 furnishes possible in-circuit voltage levels for single and double stage 12V-powered CE amplifiers for different component defects. This will help in tracking down a faulty component in an

DC VOLTAGES FOR TROUBLESHOOTING A CE AMPLIFIER

12V Vcc

BIAS COMPONENTS	V_B	V_E	V_C
NORMAL	2.0	1.5	5.8
OPEN R1	0	0	12
OPEN R2	3	2.5	2.5
OPEN R3	.7	.175	.18
OPEN R4	2.1	2.1	12
OPEN C1, C2, C3*	2.0	1.5	5.8
SHORTED C1 TO GROUND:	0	0	12
FROM NEXT STAGE:	3.25	2.7	2.7
SHORTED C2 TO GROUND:	.5	0	0
TO NEXT STAGE:	2.0	1.5	3.25
SHORTED C3	.55	0	.05
TRANSISTOR	V_B	V_E	V_C
OPEN **CB** JUNCTION	0.68	1.75	12
SHORTED **CB** JUNCTION	2.7	2.2	2.7
OPEN **EB** JUNCTION	2.1	0	12
SHORTED **EB** JUNCTION	0.1	0.1	12
SHORTED **CE** JUNCTION	2.1	2.3	2.3

ALL VOLTAGES REFERENCED TO GROUND

12V

R_1 56kΩ R_3 5kΩ V_C C_2 10μf

C_1 10μf V_B

R_2 12kΩ R_4 1.2kΩ C_3 10μf V_E

*An open C3 would cause a lower AC signal out;
an open C1 or C2 can cause a total lack of AC out.

Table 10-1. *DC voltages for troubleshooting a CE amplifier.*

ordinary class A amplifier. (Shorted and opened component values assume complete, and not partial, shorts and opens.)

10-8B. *Transistor Out-Of-Circuit Test*

If in-circuit tests indicate that the transistor may be damaged, remove the unit and perform an out-of-circuit test. Even though advanced transistor testers exist, they are rarely needed. Most tests can be performed with a basic ohmmeter. First confirm that the red, or positive, lead on the ohmmeter is actually at positive potential, as many are not. By placing the ohmmeter's leads as indicated in *Table 10-2*, with the ohm's scale set at R x 100 and, if so equipped, the low-power ohms position *not* chosen, you will quickly obtain relative good/bad indications of a transistor's function. Even though any LOW indication will vary tremendously in ohmic value from transistor type to transistor type, the HIGH readings should be almost infinite.

10-8C. *N-Channel FET Out-Of-Circuit Test*

To test for a good/bad FET, set the ohmmeter to R x 100, connect the negative lead to the gate, and the positive lead to the source or drain. The ohmmeter should show almost infinite resistance. If a medium or low resistance reading is indicated, then the device is leaking. To further test the FET, reverse these connections to forward bias the device. The ohmmeter should indicate between 500 to 1,000Ω. Set the ohmmeter to *R x 10k*, and attach the positive lead to the drain and the negative lead to the source. Hold the drain lead; then touch the gate. The forward resistance from the source to the drain should decrease.

RELATIVE INDICATIONS FOR
A GOOD NPN TRANSISTOR

E	B	C	RELATIVE OHM VALUES
+		-	HIGH
+	-		HIGH
	-	+	HIGH
-	+		LOW
-		+	LOW
	+	-	LOW

Table 10-2. *Relative indications for a good NPN transistor.*

VOM

CURRENT LIMITING
RESISTOR

DC POWER SUPPLY

VOLT CURRENT
ADJ ADJ

Figure 10-16. Test setup for
checking a Zener diode's VZ.

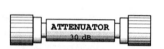

Figure 10-18. A common coaxial
in-line fixed attenuator.

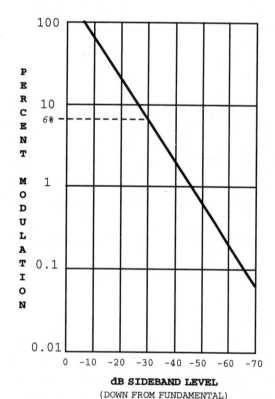

dB SIDEBAND LEVEL
(DOWN FROM FUNDAMENTAL)

Figure 10-20. Graph used to calculate AM
modulation percentage from a spectrum analyzer.

Figure 10-17. A popular step attenuator.

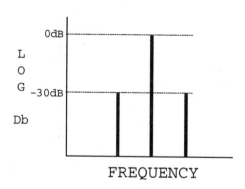

Figure 10-19. Setup for observing an RF signal
by using a pickup loop (a), or a small antenna (b).

FREQUENCY

Figure 10-21. A spectrum analyzer display
of an AM signal at 6% modulation.

10-8D. MOSFET Out-Of-Circuit Test

Attach any lead of the (low voltage) ohmmeter, which must be set to the highest ohms scale, from the gate to the source or drain. Unless infinite resistance is indicated, the MOSFET is leaky or shorted.

10-8E. Capacitor Out-Of-Circuit Test

Make sure the capacitor is discharged by shorting both leads. Using an ohmmeter, attach one lead to each side of the capacitor. This charges the capacitor through the ohmmeter's internal battery. Reverse the leads. The ohmmeter's needle should now rise quickly, then fall as the capacitor is discharged through the meter. This indicates a good capacitor. If there is a low constant resistance across the capacitor, then it is shorted — however, keep in mind that an electrolytic capacitor may have only about 0.5 MΩ of resistance and still be normal. If there is an infinite resistance reading with no rise and fall of the needle, then it is opened. (Reliable only on capacitors larger than 100 pF.)

10-8F. SCR Out-Of-Circuit Test

To test low-current SCRs for proper function attach an ohmmeter, set to the *R x 1* scale, as shown in *Figure 10-15*. Take a reading across the SCR with the polarities as shown. The resistance should be almost infinitely high. Now short the gate to the anode and the resistance should drop to under 50Ω. When the short from gate to anode is removed, the resistance reading should remain the same.

10-8G. Diode Out-Of-Circuit Test

To verify a properly functioning diode, utilize this diode test procedure. Employing an ohmmeter, set to the *R x 100* range and forward bias the diode by attaching the ohmmeter's probes to the diode's leads — but do not touch fingers to leads or probe tips (bodily resistance will affect reading). Many ohmmeters do not have a positive voltage at the plus lead, or a negative voltage at the negative — always test, then mark, leads for proper polarity. Also confirm that the ohmmeter has enough forward voltage to "turn on" the diode. When forward biased, an ohmmeter indication of approximately 600Ω is normal for silicon diodes, less for germanium, and can be appreciably more for some high voltage types. Any values that significantly vary from these amounts indicate either a shorted or opened diode. When reversed biased the resistance readings should be, for a silicon diode, almost infinite (up to hundreds of megaohms); and for germanium anywhere between 100 kΩ to 1 MΩ or more. Any values substantially less indicate a leaky or shorted diode.

10-8H. Zener Voltage Out-Of-Circuit Test

The simple diode ohmmeter test above will indicate opens, shorts and leakage, but not a Zener's proper Zener voltage (V_z). To test this Zener voltage, reverse bias and connect as per *Figure 10-16*. Supply the diode's rated Zener current by increasing the power supply's output voltage until the Zener's current rating is reached. Vary this value above and below the rated Zener current while monitoring its voltage

Figure 10-22. *Vmin and Vmax of an AM modulated waveform in the time domain.*

Figure 10-23. *Finding the approximate frequency deviation of an FM signal.*

Figure 10-24. *Single-tone AM overmodulation as seen on a spectrum analyzer.*

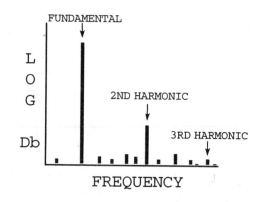

Figure 10-25. *Check at least out to the third harmonic for excessive harmonic generation.*

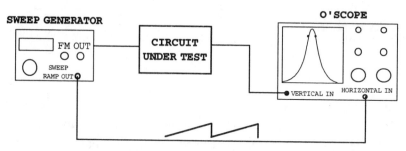

Figure 10-26. *Viewing the bandpass of an amplifier with a sweep generator and oscilloscope.*

drop and current. (Normal Zener currents are approximately 50 mA for a 1-watt, 5.1V Zener and around 20 mA for a 1-watt, 12V Zener — the higher the Zener voltage, however, the lower the Zener current that the device can withstand, given the same power rating.) If the Zener voltage remains relatively constant, and as rated, while the input voltage is varying, then the Zener is operating properly.

10-8I. AM Modulation Percentage Test

First, a word of caution: *Never place a transmitter's signal directly into either an oscilloscope's or spectrum analyzer's input.* The oscilloscope's input has an impedance of 1 MΩ, which would reflect almost all of the transmitter's power directly back into the transmitter's finals, possibly destroying them. The spectrum analyzer's front end would be destroyed by any power level over 1W entering its input. However, the spectrum analyzer can be hooked up to the transmitter's output by feeding a *properly attenuated* transmitter signal into the analyzer's input, with the amplitude of its vertical amplifier set to logarithmic. (See *Spectrum Analyzer.*) The necessary attenuation can be accomplished by using a *step attenuator* (*Figure 10-17*), which is capable of varying the attenuation in preset steps either by manually turning a knob, or setting switches, or by electronic control. If the power output of the transmitter is beyond the safe operating range of the step attenuator, which is commonly the case, then employ a fixed-value *power attenuator* in back of it — while adjusting the needed value of attenuation with the step attenuator. Small coaxial in-line attenuators are also available that can supply various levels of fixed attenuation (*Figure 10-18*) of up to 30 dB, with a maximum safe power dissipation of 25W. 2W low powered models can supply up to 60 dB attenuation.

To test the percentage of AM modulation, set the microphone gain, if so equipped, to its center position. Inject the audio modulation at a desired amplitude into either the microphone or directly into the microphone's input jack. Confirm that the output of the transmitter is appropriately terminated by an antenna, or by a dummy load capable of handling the transmitter's full output power. The percentage of AM can then be checked either with an oscilloscope or with a spectrum analyzer. When using the oscilloscope or spectrum analyzer employ an RF pickup loop or small antenna (*Figure 10-19*) placed close to the transmitter's antenna, its dummy load, or near to the transmitter's final tuned circuit. A spectrum analyzer can also have the transmitter's output fed into its 50Ω input if proper series attenuation is employed, while the oscilloscope's probe can be attached in parallel to the transmitter's load, or a small portion of the transmitter's power can be tapped off and fed into the scope through a "T" connector or line sampler. When working with the spectrum analyzer, read the signal's sideband and fundamental amplitudes on the CRT display, take the *difference*, in dBs, between the two, and use the graph of *Figure 10-20* to accurately interpret the modulation percentage. As an example, utilizing the analyzer display of *Figure 10-21*, the sidebands are shown to be exactly 30 dB down from the carrier. Glancing at *Figure 10-20* and finding the -30 dB on the lower horizontal *dB sideband level* scale, and reading up and across to the vertical *percent modulation* scale, we find the modulation percentage to be 6%. (The vertical scale is logarithmic, and not linear.) When using an oscilloscope, use the formula:

$$\frac{V\max - V\min}{V\max + V\min}$$

to easily calculate the modulation percentage.

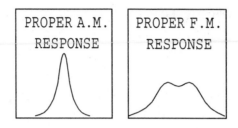

Figure 10-27. *Appropriate IF-section frequency responses.*

Figure 10-28. *Finding the passband or resonant frequency of a test circuit.*

SQUARE WAVE IN	O'SCOPE	SPECTRUM ANALYZER
GOOD		
BAD **LOW** FREQUENCY RESPONSE		
BAD **HIGH** FREQUENCY RESPONSE		

Figure 10-29. *The squarewave analysis with possible test results for an audio amplifier.*

For example, in *Figure 10-22*, if V_{max} equaled 5V and V_{min} equaled 1V, then the modulation percentage is calculated as:

$$\frac{5V-1V}{5V+1V}100 = 66.6\%$$

The oscilloscope can be a very low bandwidth type, since only the low frequency modulation envelope is of interest, if the AM output is tapped-off of a dummy load (through a coaxial "T" connector) and inserted directly into the scope's vertical-deflection plates. If a large AM transmitter is equipped with a panel mounted oscilloscope and set to display *trapezoidal* modulation patterns, these can be employed to read the modulation percentage. (See *Appendix*.)

10-8J. FM Deviation Test

To find the *approximate* deviation of an FM signal on a spectrum analyzer, locate the sideband where the amplitude begins to drop and *keeps on dropping*. (*Figure 10-23*.) The frequency spread between the carrier and this sideband is the deviation.

An *FM deviation-meter*, if available, will supply a simple, fast and accurate deviation reading.

10-8K. Harmonics and Overmodulation Test

The harmonics and excessive bandwidth created by single-tone AM transmitter overmodulation will show on a spectrum analyzer display as very high amplitude sidebands pairs (LSB and USB, both within 6 dB of the carrier), with the generated harmonics evenly spaced from each other in multiples of the *modulation* frequency. (*Figure 10-24*.) If voice or music were the baseband signal, the harmonic frequency generation would be far more complex. In addition, watch for transmitter carrier harmonic generation at least out to the third harmonic (*Figure 10-25*) by widening the frequency span on the spectrum analyzer and confirming that harmonic suppression is within specifications for the equipment under test.

If placed in the field without a spectrum analyzer, harmonic generation and suppression can still be measured with other test equipment. At some distance from the transmitter site, while utilizing a *tunable field-strength meter* with the output fed to a *dB meter*, take a dB reading of the fundamental frequency and its second harmonic by tuning the field strength meter to those frequencies and notating the results. The difference between the two readings is the amount of harmonic attenuation.

10-8L. Amplifier Bandpass Test

By the use of a *sweep generator* and oscilloscope, the bandpass of an amplifier or circuit can be determined. The setup is shown in *Figure 10-26*. As the sweep generator sweeps through a desired band of frequencies, the bandpass response of an amplifier or tuned circuit is displayed on the scope's CRT. *Figure 10-27* shows typical AM and wideband FM IF-section receiver responses. The bandwidth is measured as the frequency values between the half power points (or at the points where the waveform's voltage falls to 70.7 percent of its peak amplitude) by counting the horizontal graticule divisions be-

Figure 10-30. *Phase measurement between two voltages.*

Figure 10-31. *Setup to check receiver sensitivity.*

Figure 10-32. *Setup for SSB PEP test.*

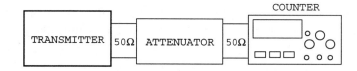

Figure 10-33. *Setup for AM/FM transmitter frequency test.*

Figure 10-34. *A handmade spectrum analyzer probe.*

tween these two points. An alternate setup is shown in *Figure 10-28*. Utilizing an oscilloscope's internal trigger as well as a simple function generator that can be manually tuned from just below to just above the bandpass of the circuit, similar results can be obtained. When the point of resonance of the circuit under test is reached, the oscilloscope will display a peak indication of the frequency generator's frequency. By measuring the frequency on either side of this peak at the calculated half-power points, an approximate measurement of the circuit's bandwidth can be found.

Another amplifier test, referred to as *squarewave analysis*, is employed to discover if an audio amplifier has the proper linear frequency response within design parameters. Begin by feeding a 1 kHz squarewave pulse train into the input to the amplifier under test, at an amplitude that will not introduce distortion. Use an oscilloscope or spectrum analyzer placed at the output of the appropriately terminated amplifier and compare the waveforms of *Figure 10-29*. Since a squarewave produces odd harmonics, feeding a squarewave into the amplifier's input indicates any possible frequency distortion. A relatively clean output verifies that the amplifier is passing at least up to the ninth harmonic without excessive attenuation or overamplification of certain frequencies.

10-8M. Phase Angle Comparison Test

To compare two phase angles, both must have the same frequency and be sinusoidal waves. Using a dual channel oscilloscope, feed each signal into a separate scope channel. Measure the time between the two signals (T_2, or the time between points A and B, *Figure 10-30*). Now, measure the time period of one complete waveform, or T_1. Apply the formula $T_2 / T_1 x 360^o$ to obtain the phase angle between the two signals.

10-8N. Resistance In-Circuit Test

Switch off all power to the circuit under test and discharge all capacitors. When making in-circuit resistance tests confirm that all semiconductors will be reversed biased by the ohmmeter voltage — or false readings will be obtained if the PN junction is forward biased "on". As an alternative, use the low-power setting on an appropriately equipped ohmmeter so that any forward biased junction will not be turned "on". Also, utilize only the *R x 100* or *R x 10* ranges on standard ohmmeters when testing semiconductor circuits, as this lowers the ohmmeter's short-circuit current. This will prevent any possible transistor damage if its PN junction should be biased "on" unintentionally.

10-8O. Receiver Sensitivity Test

To check for correct receiver sensitivity at its rated S/N ratio, inject an RF generator's signal into the receiver's input, preferably by removing the antenna and directly attaching a coax cable from the generator. (*Figure 10-31*.) This generator's signal output should be at the receiver's rated minimum sensitivity level (frequently 1 uV or, if the receiver is suspected of having lower than rated sensitivity, increase the initial voltage input), and at the exact dial frequency along with 30% 1 kHz modulation. With the receiver unsquelched and the volume at a reasonable amplitude, read the output, in dB, at the properly terminated speaker output with a voltmeter set to the decibel scale. Then remove the modulation at the

RF generator and read the output in dB (the receiver's *no signal* condition). The difference is the S/N ratio, and is ordinarily at least 10 dB. In this case, the specification would be *sensitivity 1 uV/10 dB (S+N)/N*. In other words, the 1 uV signal is 10 dB above the noise floor. Certain obligatory audio output powers (such as 1 watt) may be specified while measuring this sensitivity and S/N ratio.

10-8P. Receiver Frequency Response Test

A frequency response test is performed on a communications receiver by hooking a modulated generator to the input of the receiver, while monitoring the output at the speaker with a DMM, oscilloscope, or audio frequency analyzer. The modulation at the generator is varied between 300 Hz and 3 kHz. Up to a 3 dB drop or less in audio output at the speaker between these two frequencies is considered acceptable for most receivers.

10-8Q. SSB PEP Transmitter Test

Attach an oscilloscope across (in parallel with) a properly terminated transmitter's antenna output. (*Figure 10-32*.) Feed equal-in-amplitude two-tone audio signals, usually at 500 Hz and 2,400 Hz, directly into the microphone or the microphone's panel input. (SSB must be modulated to produce any transmitter output.) Increase, equally, these audio amplitudes until the output SSB signal is just under flattopping on the scope. Then obtain the PEP reading of the displayed waveform by using the formula $PEP = V_{RMS}^2 / R$; with the R being the value of the termination (ordinarily 50Ω).

If a peak-reading RF wattmeter is available, this can be employed to indicate the peak (maximum) envelope power when voice is used to modulate the transmitter, since a standard wattmeter will only indicate the average power of the signal, which may be as low as one-fourth of the PEP value over time. In other words, the peak-reading wattmeter senses the highest maximum (peak) value of the modulated SSB waveform, and displays this as the PEP reading — while not averaging in the lower amplitude peaks, which would significantly decrease the reading and give no indication of the power in the peaks of the modulated RF signal, but only the average power over time.

10-8R. SSB Frequency Test

To check the frequency of a SSB transmitter, modulate the transmitter with a 1 kHz tone injected into the microphone, or directly into its MIC input jack, and make a properly attenuated frequency counter measurement of the transmitter's output (or place a pickup loop or small antenna into the BNC connector of the frequency counter to make an indirect but safe measurement). The USB/LSB switch should be set to USB, with any *TX RIT* (or *TX clarifier*) control switched out or set to "0". Subtract 1 kHz from the indicated reading. The resultant is the f_r of the transmitter.

10-8S. AM/FM Output Power Test

Place a wattmeter in-line between the transmitter and its antenna or dummy load, or place an oscilloscope across a proper noninductive output resistance, such as a dummy load. Switch to transmit and read the wattmeter indication or, if working with the scope, use the formula $P = V^2_{RMS}/R$ to calculate the transmitter output power. Do not modulate while the reading is being taken.

10-8T. AM/FM Transmitter Frequency Test

Setup as in *Figure 10-33*, confirming that the attenuator is reducing the power to the counter's input to safe levels. However, verify that there is adequate power to acquire an accurate reading — or place an antenna or pickup loop on the frequency counter's BNC input while the transmitter is feeding an antenna or dummy load. Switch to transmit with zero modulation applied. Read the frequency indication on the counter display.

10-8U. Ohmmeter Click Test

The ohmmeter click test can be used to quickly troubleshoot a communications receiver, in the field, with a simple VOM:

1. Power on the receiver.
2. Do not ground the negative VOM lead.
3. Set the VOM to the ohmmeter's highest range.
4. Touch the positive lead to the collector or base of an amplifier stage for an instant. If a click is heard at the receiver's speaker, then the stage or stages all the way to the speaker are good.
5. Work back from the last audio stage until no click is heard at the receiver's speaker. That stage is bad.

10-9. Test Equipment

Test equipment, such as multimeters and oscilloscopes, are standard on any test bench, while spectrum analyzers, frequency counters, frequency generators, and logic probes are considered a little more specialized. These pieces of test equipment, and the knowledge to effectively operate them, are now regarded as mandatory in the field of RF.

10-9A. RF Probes

A word about RF probes. Shielded cable passive attenuator *low-capacitance probes*, usually found on oscilloscopes, are very common and are built to have the least effect on the signal to be measured. They increase the oscilloscope's input impedance and decrease the scope's input capacitance, and are designed not to distort the input signal within the probe's rated bandwidth. Low-capacitance probes are important to lessen shunt capacitance (which would attenuate high frequency signal amplitudes), reduce pickup of stray fields (which would create interference), diminish capacitive loading (which would cause distortion of non-sine waves), and decrease circuit loading (which would cause incorrect voltage indications and create circuit instability). *RF* and *demodulator probes* are operated with VOMs or oscilloscopes and are built to convert the RF signal into either an RMS or a peak-reading DC voltage (depend-

Figure 10-35. A no-frills spectrum analyzer.

ing on model), allowing the low frequency VOM or scope to furnish relatively accurate voltage indications. (The demodulator probe produces a steady DC voltage when RF is present, and AC when AM is present.) Some are reliable up to 0.25 GHz. Keep in mind that many RF and demodulator probes do not necessarily give the true voltage of the circuit under test, but only a relative indication. *Active FET probes* are available and are capable of very high bandwidth operation (up to 2.5 GHz), and very high prices (up to \$3,500). They have very low input capacitances (0.6 pF), and are designed for oscilloscopes. *Passive divider probes* are also available that can function up to 6 GHz with extremely low input shunt capacitances of around 0.25 pF. Input adapters are also available to convert the 50Ω input impedance common to spectrum analyzers into 1 MΩ, allowing the analyzer the capability to carry out high-impedance (1 MΩ) measurements, similar to an oscilloscope. However, they are commonly good up to only 100 MHz and cost about \$1,000. Higher frequency probes that supply 1 MΩ of input resistance and a bandwidth of up to 3 GHz are also available for spectrum analyzers for about \$2,500. These devices are utilized because the spectrum analyzer's low 50Ω input impedance would excessively load down the circuit-under-test while probing circuits during troubleshooting. The spectrum analyzer probes can only be used in low powered circuits, such as receivers and the low powered sections of a transmitter. They are damaged above relatively low input voltages (1.5 to $2V_{RMS}$), and are ordinarily equipped with a 10:1 voltage divider to increase this maximum voltage to about $20V_{RMS}$. If such spectrum analyzer probes or adapters are not available, a simple probe can be manufactured (*Figure 10-34*) that will exhibit to any circuit-under-test a 5 kΩ load for about \$2.

Caution: *If high DC levels are present during probing*, use a blocking capacitor in series with the probes (if the probe is not already so equipped) to protect the signal generator, spectrum analyzer, or frequency counter from damage.

10-9B. The Multimeter

Multimeters, as with all test equipment, are available with varying features, qualities and prices. A digital multimeter (DMM) is used far more than the older analog type of VOM. The DMM is capable of measuring AC and DC voltages and currents, as well as resistance values, with high accuracy. Many are equipped with auto ranging, dBm measurement abilities, a diode tester, a continuity buzzer, and even transistor gain and capacitance measurement capabilities. The analog VOM type is, however, superior to the DMM for keeping track of rapidly changing current and voltage level trends. Also, when working near an active transmitter, DMMs can be severely influenced by the RF, causing erratic operation — under these circumstances, an analog model must then be used.

The input impedance of a multimeter is measured in *Ω/V* of sensitivity. If using an analog meter, do not settle for less than 50,000 *Ω/V* so that loading is kept down to negligible levels in most circuits. When working with a digital model, or a FET-input analog model, any good quality unit will have 10 MΩ input impedance or better.

Keep in mind that most VOMs will only give accurate readings when the AC voltage being measured is at 20 kHz and under for better analog units, or up to 100 kHz for most quality DMMs.

10-9C. The Spectrum Analyzer

In RF, the ability to operate and interpret the results of a spectrum analyzer is essential. While an oscilloscope is ordinarily used more often, there are occasions when a time-domain reading can not give all the information necessary. Since a spectrum analyzer reads in the frequency-domain (a spectrum analyzer displays a range of frequencies-versus-amplitude), it is capable of presenting all the frequency components that shape a wave, as well as display an entire spectrum of diverse signals. A signal that seems clean on an oscilloscope may appear to be anything but when observed on a spectrum analyzer's CRT, since harmonics must be at least 20 dB or less from the fundamental to be seen reliably as distortion of a sinewave on the oscilloscope's CRT. Also, very high frequency oscilloscopes are expensive and quite rare, while the *average* spectrum analyzer is good to one gigahertz.

Different designs for spectrum analyzers have been used, but the most common is the superheterodyne class. This type of spectrum analyzer can be visualized as nothing more than a normal superheterodyne receiver that visually displays all frequencies within its passband, rather than outputting an audio signal, by sweeping, or tuning, its local oscillator through a range of frequencies. The amplitudes of these frequencies are displayed in the vertical, while the scanned frequencies are displayed in the horizontal (see *Figure 10-24*), with the highest frequency to the right of the CRT and the lowest to the left. The amplitude of a signal is normally shown in a logarithmic fashion, in decibels. If all incoming signals were displayed in a linear manner, the usable dynamic response of the analyzer would be quite low. Consider that if input signals differed by just 60 dB, one incoming voltage would differ by the other by one thousand times; small signals would get lost, while large signals would overdrive the display. Any meaningful comparison between the two signals would be almost impossible.

There are many exotic uses for the spectrum analyzer, but the application that the technician is most concerned with is the spurious content of a transmitted waveform. A transmitter must be tuned for the highest spectral purity attainable, since harmonics, IMD, and parasitic output must be kept to a minimum — or severe interference with other stations will occur (as well as creating inefficient transmitter operation).

To operate a basic spectrum analyzer, refer to *Figure 10-35*. **First, a warning**: *Any high input levels, whether they are DC or AC, will instantly destroy the analyzer's front end.* External attenuation *must* be employed for any possible high amplitude signals entering the analyzer's front end. (*Figure 10-36*, and see *Attenuators*.) Low-powered transmitters (one watt or less) may be attenuated solely with the analyzer's *INPUT ATTN* (or *input level* or *reference level*) knob set to +30 dBm initially. If working with such low powered transmitters, the transmitter and analyzer can be connected directly through a coax cable. (The transmitter would typically take a BNC or PL-259 connector, while the analyzer would normally take an "N" connector, though some are outfitted with BNCs.) The *INPUT ATTN* knob, as the name indicates, will attenuate the input signal — normally in calibrated discrete steps of 10 dB, with a fine adjustment in 1 dB steps. In most analyzers, this input attenuation value is now our starting reference point for all amplitude measurements which, on a spectrum analyzer, is the very top line of the CRT display, and is referred to as the *reference level*. For example, if the input attenuation is set to 0 dB, then the top line of the display is at 0 dB — any input signal is now referenced to that line, and anything below that line is of a negative dB value. Obviously, due to the analyzer's sensitivity, it would be judicious to start with the maximum attenuation first, and work down, while viewing the entire frequency spectrum by using the *SCAN WIDTH* knob at its widest setting (generally marked as *Max*). This is to insure that there are no high amplitude signals further up or further down in the frequency spectrum that can destroy the analyzer's front end. The *SCAN WIDTH* (or *frequency span*) knob adjusts the frequency range of the sweep across the frequency spectrum to be displayed on the CRT. In the case of the analyzer of *Figure 10-35*, anywhere from a width of 100 Hz to 500 MHz can be scanned, or swept. Many analyzers, though, are calibrated in *scan width-per-division* (or *freq. span/DIV*). To obtain the entire swept spectrum range of the display, simply multiply the number of horizontal divisions (ten) by the number across from the scan width/DIV dial indication. The internal bandwidth of the analyzer itself is adjusted by the smaller inset *RESOLUTION* knob, which alters the filter selectivity (bandwidth) of the IF stages of the analyzer. This setting is used when two sinusoidal signals are very close together in frequency — the narrower the setting, the easier it is to separate these two signals. An excessively narrow resolution setting, however, will create erratic responses to be displayed on the CRT, since any analyzer instability will create a false or distorted display. As a rule, it should be set to *coupled*. The *FREQ TUNING* (or *center frequency*) knob allows the technician to center the frequency of interest on the display (similar to a radio's tuning knob), while choosing the width of the scan on either side of this frequency with the scan width knob. The vertical (amplitude) range of the display is adjustable by the *LOG/LIN* switch, indicating 10 dB/DIV, 2 dB/DIV, and LIN (linear). By switching to the 10 dB/DIV range, each vertical division will now equal a value of 10 dB. This setting is used for normal measurements with signals that are expected to have high dynamic range and amplitude. The *2 dB/DIV* setting is for more precise, low amplitude measurements. Switching to LIN will give the amplitude an (ordinarily) RMS linear response for reading very low amplitude values, with a low dynamic range, of any input signals. The *BASELINE CLIPPER* control merely "clips-off" the bottom of a signal to remove any annoying noise (grass) present in the lower part of the display. The *LOW FREQ FILTER* (or *variable video filter*) is a low-pass filter that smooths out the display and rids it of excessive noise hash. On many units an adjustable *SWEEP* (or *time/DIV*) knob is also available to regulate the sweep speed of the trace across the CRT. If the sweep time is set at too fast a speed, the apparent amplitude of the signal will decrease and a shift will occur in the signal's apparent frequency. As needed, set the sweep knob for a flicker-free and stable display or, typically, switch to *auto*. Generally, most of the setting's values made by the operator are arrayed at the top of the CRT for easy and quick reference.

If taking any absolute amplitude measurements with a spectrum analyzer, such as the amplitude of a carrier, rather than *ratios* between a signal and its harmonics, it must be remembered to compensate for the amount of external attenuation employed, and adjust your amplitude readings accordingly. Also, since most RF spectrum analyzers have a 50Ω input impedance, there will be an impedance mismatch between a 75Ω output transmitter and the 50Ω input of the analyzer, with the normal loss of power and inaccurate amplitude readings. (There are impedance matching couplers available, with an internal DC blocking capacitor, to compensate for this mismatch.)

To sustain absolute amplitude calibration of an input signal on the analyzer's CRT screen (so that any input signal has a definite and unchanging value), no matter what the analyzer's input attenuation setting may be set for, two methods are commonly employed to compensate. *The first*: when setting the input

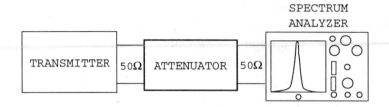

Figure 10-36. Proper spectrum analyzer setup for testing a transmitter.

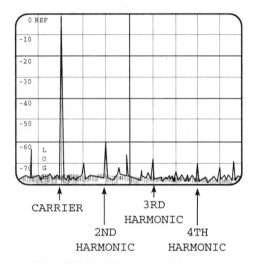

Figure 10-37. An unmodulated carrier with properly attenuated harmonics.

Figure 10-38. A good single-tone AM waveform.

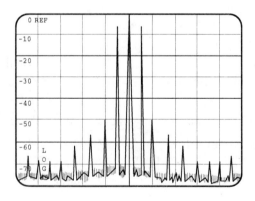

Figure 10-39. A single-tone AM overmodulation waveform.

Figure 10-40. A superior two-tone SSB signal.

attenuation knob, the indicated reference level (RL, the value of the top line of the CRT's graticule) will automatically change value to compensate for this variation in input signal strength. As the analyzer's input attenuation is increased, the displayed signal will decrease in amplitude, while the reference level's value at the top of the graticule will increase in value (in step with the panel mounted input attenuation knob's setting). This allows the operator to calculate the signal's true power by starting at the reference level, and counting the divisions down to the peak of the signal. *The second*: with some newer analyzers, as the input attenuation is increased, the displayed signal will always remain at the same amplitude (the analyzer's IF amplifier gain increases to compensate for the increase in input attenuation), while the reference level also remains unchanged. This method automatically compensates for any input attenuator settings by allowing an input signal to remain at the same relative location on the CRT's graticule, and in relation to the reference level, no matter what input attenuation settings are employed. In other words, both of these methods assure that if a -10 dB signal is placed at the input to the analyzer, then the CRT will allow the operator to observe instantly that the amplitude of this signal is -10 dB, whether this signal is heavily attenuated by the input attenuator or not.

A movable *marker* indicated on the CRT is also available on some analyzers. This marker can be placed, ordinarily by arrow cursor buttons on the analyzer's front panel, at any spot along the CRT. It enables the operator to quickly read the frequency and amplitude at any desired location on the CRT, which is then displayed either on the CRT face or on a separate panel meter.

Up-converting superheterodyne analyzers have what is referred to as *local oscillator feedthrough*. The internally generated first LO signal reaches the analyzer's display and causes an indication that appears to be a high amplitude signal. Since this signal is always at 0 Hz, it is a useful indicator of where the technician is looking in the frequency spectra, as well as allowing more accurate evaluation of low frequency signals, since we now have a precise reference point from which to begin measurements.

Although different FCC rules apply as to the maximum spurious emissions allowed to be generated — depending on the transmitter's model year, frequencies generated, output power levels, etc., the following examples will give you a starting idea of the spectrum analyzer displays that will commonly be encountered. It is recommended that the *scan-width* of the analyzer be set wide at first to catch any high amplitude harmonics, or other spurious responses, that would cause interference to other stations from the transmitter under test. The *Figure 10-37* display of an unmodulated carrier from an AM transmitter shows a good spectrum out to the third harmonic, with the second harmonic 60 dB below the fundamental. *Figure 10-38* illustrates a much narrower analyzer sweep of a very good AM signal modulated with a single 1 kHz tone at approximately 50% modulation. (The analyzer's amplitude is set to *Log*.) The signal shows that the sidebands have equal amplitude, indicating no *incidental FM*, and all harmonics of the original modulating waveform (harmonic sidebands), are at least 60 dB below the fundamental, with no *parasitics* present. *Figure 10-39* presents a poor AM display with *overmodulation* causing severe spurious responses. *Figure 10-40* presents an excellent SSB two-tone signal with the carrier and the opposite sideband, in this case the LSB, properly suppressed, with little intermodulation distortion (IMD). An inadequate two-tone SSB display showing extensive IMDs is exhibited in *Figure 10-41*. IMDs should, as a general rule, be at least 40 dB below the PEP output. The display of *Figure 10-42* demonstrates an unmodulated FM signal with all carrier harmonics well under the legal limit. *Figure 10-43*

ORIGINAL TWO-TONE
USB SIGNAL

Figure 10-41. A two-tone SSB
signal with massive IMD levels.

CARRIER

Figure 10-42. A great
unmodulated FM carrier with little
harmonics present.

Figure 10-43. A single-tone
FM signal showing sidebands.

displays a good FM single-tone modulated spectral exhibit. Depending on the modulation index being used, FM displays can vary drastically from each other and yet still be completely acceptable. (See *FM Fundamentals*.)

Short instructions for use:

1. Warm up all equipment and the unit-under-test (in this case a transmitter).
2. Place attenuators, or use a line sampler, between analyzer and transmitter (EX: At least 20 dB attenuation is needed for a 100 watt transmitter).
3. Set analyzer preliminarily as follows:
    a.   INPUT ATTENUATION to maximum (+30 dBm).
    b. SWEEP SPEED to auto.
    c. LOG/LIN switch to 10 dB/DIV.
    d. SCAN WIDTH to maximum (Max).
    e. RESOLUTION to coupled.
    f. BASELINE CLIPPER completely counterclockwise.
    g. VIDEO FILTER (LOW FREQ FILTER) off.
    h. INTENSITY, FOCUS, SCALE ILLUMINATION knobs adjusted as desired.
4. Switch on transmitter to transmit.
5. Adjust analyzer's INPUT ATTENUATION for desired signal amplitude.
6. Center signal of interest with FREQ TUNING knob.
7. Narrow SCAN WIDTH as desired to view signal or its harmonics.
8. Engage BASELINE CLIPPER and/or LOW FREQ FILTER as desired.
9. Take measurements of signal and any spurious responses.

10-9D. The Oscilloscope

The oscilloscope is an indispensable tool for all who work in electronics. It is a device that allows the operator to view the shape of any type of waveform in the time domain, as well as measure its amplitude, phase, and frequency. Any artifacts (anomalies) present in the waveform or distortions of the waveform will be quickly detected.

The oscilloscope of *Figure 10-44* is a basic no-frills dual-trace oscilloscope. The *TIMEBASE* (the *horizontal amplifier*) adjusts the sweep speed, either in milliseconds or microseconds per division, of the electron beam that traces out the signal being viewed on the face of the CRT. This horizontal axis exhibits the signal in the time domain. The faster the sweep, the smaller the time slice that is being displayed, thus allowing a higher frequency to be viewed. The *CH 1* and *CH 2* knobs adjust the sensitivity of the vertical amplifiers of each channel's input signal. The vertical axis displays the amplitude of the signal, typically in volts and mV. Since the values given around each knob decrease as this knob is rotated in the clockwise direction, any input can be increased or decreased in amplitude to adjust the height of the signal on the CRT. These values, as in the horizontal amplifier, are per division. Thus any value read must be multiplied by the amount of graticule divisions the signal's maximum positive and negative amplitudes reach for a peak-to-peak voltage reading. The *AUTO/NORM* button can be switched

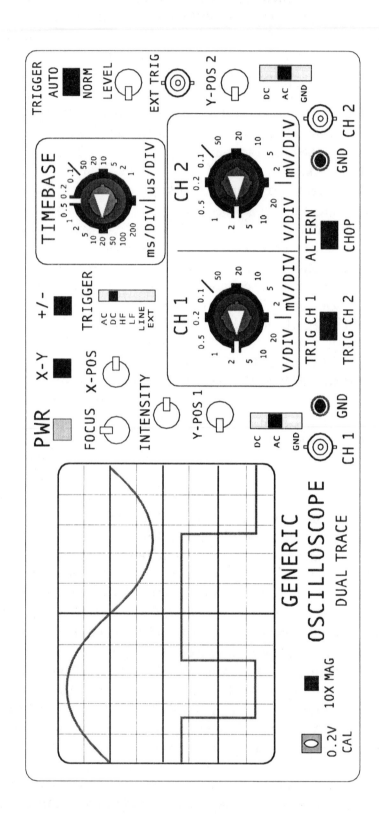

Figure 10-44. The dual-trace oscilloscope.

to *AUTO* to set the trigger to automatically begin a sweep at the same relative part of the waveform each time (such as when an input signal crosses zero volts, going from negative to positive), allowing for a stable display that does not float across the CRT. If the button is pushed to *NORM*, triggering must be manually adjusted by the *LEVEL* knob, which allows the operator to adjust the triggering to occur at any point on a waveform. This is normally employed with complex waveforms that the auto trigger mode cannot handle. Triggering on the positive or negative-going edge of a waveform can be chosen by depressing the -/+ button. The *TRIGGER* selector switches in different filters in-line with the trigger circuits to supply more reliable triggering at different frequencies. The AC setting is utilized to block any DC components, and can be employed for most signals; the DC setting will pass DC or very slow AC, as well as being the recommended setting for most digital input signals; *HF* filters out low frequency fluctuations; *LF* filters out high frequency interference and noise; *LINE* is adopted to supply triggering at power line frequencies; *EXT.* works in conjunction with any trigger signal at the *EXT. TRIGGER BNC* input. *FOCUS* and *INTENSITY* adjust the CRTs electron beam (trace), while the slider switches marked *DC, AC, GND* adjust the input coupling for each channel into the vertical amplifiers. With *DC* coupling both DC and AC signals are passed to the vertical inputs; *AC* coupling blocks the DC, as well as low frequency alternating current, and passes the AC; *GND* places chassis ground at the input, allowing for a reliable no-input reference to adjust the traces. *X-POS* adjusts the position of both horizontal traces on the face of the CRT; the two *Y-POS* knobs adjusts the position of each vertical trace. *CH 1* and *CH 2 BNC* connectors accept the standard 10:1 (or 100:1) attenuator probes (*Figure 10-45*), or any other suitable probe. (See *Probes.*) Through the probe the input signal is fed into the vertical amplifiers. The *10x MAG* button magnifies a portion of the signal by ten times for closer observation of the waveform. The *0.2V CAL* connector outputs a $0.2V_{P-P}$ 1 kHz square wave for compensating (adjusting) the 10:1 attenuator probes. (Probes may have both a high and low frequency adjustment or merely a single adjustment — either way, the RC time constant of the probe must be adjusted for accurate, undistorted measurements. Simply turn the internal probe trimmer until a perfect square waveform is obtained on the CRT, with no droop or overshoot.) The *TRIG CH 1/TRIG CH 2* push-button chooses between channel 1 or channel 2 for triggering (when using both channels). *ALTERN*, also used when working with both channels, selects between *alternately* completely sweeping the 1st channel from left to right, then sweeping the second channel completely, then the first, etc. If the sweep speed must be very slow, then the first sweep will begin to fade before it can be refreshed by a new pass of the electron beam. When this occurs, *CHOP* mode should be adopted. Each trace is now chopped up into many equal sections; the beam switches between the two traces at a very high speed, not allowing either to fade. The *X-Y* measurement button is pressed to display Lissajous patterns (which is an old method for comparing phase or frequency between two similar signals).

Even without a spectrum analyzer, harmonic generation can be roughly detected with a basic oscilloscope, but only if the harmonics are at relatively high levels and within the bandwidth of the scope. (The harmonics of a waveform are, of course, much higher than the fundamental signal being measured.) If the sinusoidal wave, such as the (unmodulated) carrier of a transmitter, is deformed in any way (*Figure 10-46*), then harmonic, and possibly intermodulation distortion, products are present in the signal.

Apart from having the ability to view the condition of a signal in the time-domain, the oscilloscope gives the technician the capability to measure the amplitude and frequency of a signal. Since oscilloscopes

Figure 10-45. *Disassembled oscilloscope attenuator probes showing adjustment points.*

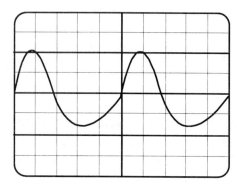

Figure 10-46. *Distorted sine wave, indicating possible harmonics.*

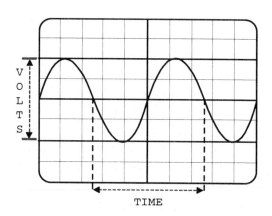

Figure 10-47. *Frequency and amplitude measurements with an oscilloscope.*

Figure 10-48. *An RF SWR/wattmeter.*

naturally display peak and peak-to-peak voltages, any RMS measurements must be calculated. Also, because 10:1 attenuator probes are typically used to decrease loading on the circuit under test, any amplitude measurements must be multiplied by ten (some scopes have a setting on the vertical amplifier knob that will take this attenuation into account). To measure the amplitude of an input signal, find the number of vertical divisions between the positive and negative peaks (*Figure 10-47*), multiply this by the volts/DIV setting. Then, if operating a 10:1 probe, multiply the voltage obtained by 10. This is the peak-to-peak amplitude of the signal. To find the RMS voltage divide by two to obtain the peak voltage, then multiply the peak reading by 0.707. Frequency can also be *approximately* determined by multiplying the number of horizontal divisions from the beginning to the end of a single wave by the time-base setting, which supplies the time period of one wavelength of the input signal. Then, use the formula $f = 1/t$ to obtain the signal's frequency.

Short instructions for use:

1. Warm up all equipment, as well as the unit-under-test.
2. Set the scope preliminarily as follows:
   a. V/DIV (vertical amplifier) fully counterclockwise.
   b. Adjust FOCUS and INTENSITY as desired.
   c. +/- set to +, TRIGGER to AC and AUTO.
   d. Input COUPLING to GND, set X and Y POSITION, switch COUPLING to AC (use DC coupling for non-sine waves or distortion can result).
   e. Confirm that 10x MAG and X-Y buttons are not engaged.
   f. TIMEBASE to center position.
3. Place the *ground clip* of the properly compensated 10:1 attenuator probe on case or chassis of unit-under-test (chassis must not be hot — use isolation transformer if so).
4. Probe the test points and check for proper amplitudes and waveforms while adjusting the TIMEBASE, VERTICAL AMPLIFIER, TRIGGER filtering, and input coupling as desired.
5. For two-channel operation, follow appropriate instructions above for the new channel, set switches to TRIG CH 1 or TRIG CH 2 and ALTERNATE or CHOP, as desired.

10-9E. The SWR/Wattmeter

Rather than manipulating Ohm's law to determine to power output of a transmitter, such as V^2_{RMS}/R if into a pure resistance, a far easier and faster method is to utilize the wattmeter, or the combination SWR/wattmeter. SWR/wattmeters measure the reflected and forward power, as well as the SWR. The unit is placed in series between the transmitter and its transmission line. While there are different types of wattmeters on the market, the example in *Figure 10-48* is referred to as a *directional single-needle* type. *Cross needle* meters are common also, allowing the operator to read both forward and reflected power, as well as SWR, at a single glance. These meters have two needles and separate forward power, reflected power, and SWR scales. One needle indicates reflected power, the other points to the forward power value, and the area where to two needles cross is the SWR.

Figure 10-49. *An RF frequency counter.*

Figure 10-50. *An RF function generator.*

To operate the single needle SWR/wattmeter of *Figure 10-48*, the transmitter is connected at the IN (or TRANS) port, with the antenna coupled to OUT (or ANT) port. To obtain a power measurement, the *POWER/SWR* button is set to *POWER*, the *HIGH/LOW* button is set to the appropriate power range expected out of the transmitter (choose HIGH for transmitters up to 2,000 watts; LOW for transmitters up to 200 watts), and the *FWD/REF* button is set to *FWD*. The AM or FM transmitter is then keyed on, indicating on the appropriate high or low *WATTS* scale the RMS output power of the transmitter. A wattmeter that is capable of reading the PEP is needed for SSB, since a SSB transmitter has no carrier and must be modulated to obtain an output signal. (See *SSB PEP Transmitter Test.*) To read the SWR present on the transmission line, switch *POWER/SWR* to *SWR*, switch *FWD/REF* to *FWD*, transmit and, at the same time, and use the *CALIBRATION* knob to set the needle to *CAL* (calibration). Cease transmission. Switch the *FWD/REF* to *REF*, and transmit. The needle will point to your standing wave ratio on the SWR scale. An indication of 1.1:1 to 1.5:1 is considered excellent, a 1.6:1 to 2:1 is acceptable, while anything over 2:1 should be looked into by checking the antenna, the transmission line, and their connections for broken or damaged parts.

10-9F. The Frequency Counter

RF frequency counters are available from the very simple, all the way up to models with every bell and whistle imaginable. The highest frequency range and the counter's accuracy are also tremendously variable. However, in virtually all modern applications, an inaccurate frequency counter has little more use than as a paper weight. The type in *Figure 10-49* has all the functions needed in 99% of real-world applications.

It is always vital to attenuate signals from a transmitter into a frequency counter. Excessively powerful signals can damage the unit, even though most are equipped with Zener input protection. The maximum power that a counter can accept goes down as the frequency is increased. All professional counters come with a derating curve with these maximum input values graphed. Typically, a maximum safe input voltage will be around 150V at 1 MHz. A very safe method to obtain a frequency reading from a transmitter is to attach either a "rubber ducky" or a small telescoping antenna to the counter's BNC input. Then, key the transmitter into an antenna or dummy load and take the measurement.

To operate a standard RF frequency counter, view *Figure 10-49*. The *LOW-PASS* button switches a low-pass filter into the counter's input when taking readings of low frequency signals (under about 100 kHz). This removes any high frequency components that may give an erroneous or unstable reading. The *10x ATTENUATOR* attenuates any input signals by ten times to save the front end from destruction from high amplitude signals. The three *FUNCTION* switches are pressed to read the *FREQUENCY* of an input signal, the *PERIOD* of a waveform, or the number of *EVENTS* that occur. The *RESET* button is used to clear the display and start the count again. *TRIGGER LEVEL* allows the operator to control where on the waveform triggering takes place. Only adjust if proper triggering does not occur when in the "0" (or *NORM*) position. The device is used by first warming up all equipment as well as the unit-under-test. For correct measurement accuracy the transmitter should be terminated in its proper output impedance, with the suitably attenuated transmitter signal then inserted into the input of the counter. (See *Figure 10-33*). If the transmitter is of sufficiently low power, a direct coaxial connection from the

Figure 10-51. *A common logic probe for troubleshooting digital circuits.*

Figure 10-52. *Proper soldering technique: Heat pad and lead (a); melt solder against pad and lead (b).*

Figure 10-53. *Removal of a two-terminal SMD.*

transmitter into the counter can be used. Either way, switch the *LOW-PASS* filter off and the *10X ATTENUATOR* on, press the *FUNCTION's FREQUENCY* button, and set the *TRIGGER LEVEL* to "0". Key the transmitter. If the input frequency is above that which the frequency counter is designed to measure, a red *OVER* indicator will light.

10-9G. The Function Generator

The function generator (*Figure 10-50*) is also an irreplaceable piece of test equipment for troubleshooting, aligning, and maintaining electronic equipment all the way from audio to RF. A function generator is a *signal generator* that can output different waveforms. It is used extensively for signal injection and signal tracing in electronic devices to check for improperly operating circuits and to measure sensitivity, selectivity, bandwidth, etc.

The *FUNCTION* section has the controls used to choose the type of output waveform desired. In the generator of *Figure 10-50* we have a choice of outputting a *SINE* wave, a *SQUARE* wave, a *TRIANGLE* wave, or *PULSE* waveforms (employed for digital circuits). *PULSE* levels are further selectable to be compatible with TTL (5V) or adjustable for CMOS circuits by the *CMOS AMPLITUDE* knob. This *PULSE* output is taken from the *PULSE OUT BNC*. The *AMPLITUDE* knob sets the level of the AC output voltage, while the *DC OFFSET* adjusts the DC offset level from 0V DC to some negative or positive value. The *DUTY CYCLE* of complex waves are capable of adjustments from 1:1 to 10:1. The *RANGE* buttons are used to choose the appropriate frequency ranges. Within the range chosen, the *TUNING* knob is varied to output the desired frequency, as indicated in the LCD or LED display. (Display readings on most generators should not be considered as accurate enough for most uses — employ a frequency counter to confirm the generator's display reading.) Since some type of modulation is needed for many RF tests, a *EXT MOD/INT MOD* switch is included. When *INT MOD* is chosen, the generator outputs a signal with internally generated 1 kHz amplitude or frequency modulation, depending on the position of the *AM/FM INT MOD* button. When *EXT MOD* is selected, a source of external audio frequency modulation can be inputted into the *EXT MOD BNC*.

To use, warm up all equipment and the unit-under-test. Set *AMPLITUDE* almost completely counterclockwise (to minimum), set *RANGE* to the proper value, place *DC OFFSET* to "0", *DUTY CYCLE* to 50%, set *FUNCTION* to the desired output waveform, and adjust the *TUNING* knob to the preferred frequency. Connect a coaxial cable between the *OUTPUT BNC* and the input of the unit-under-test; or attach a rubber duck or other type of antenna to the generator's *OUTPUT*; or attach a probe from the generator to the point of injection of the unit-under-test — whichever is appropriate. Adjust *AMPLITUDE* for proper output, choose *EXT MOD* or *INT MOD* or no modulation. If *INT MOD* is chosen, then select either the internal FM or AM modulation.

Keep in mind when changing the power output of older or low cost RF generators that the output signal may be pulled off frequency, necessitating resetting.

Figure 10-54. *Soldering two-terminal SMDs to a PCB: (a) soldering first terminal; (b) soldering second terminal.*

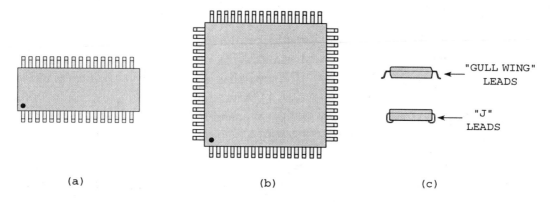

Figure 10-55. *Two IC SMDs: SOIC (a); QFP (b). Gull wing and "J" lead type SMDs (c).*

Figure 10-56. *A sniffer probe.*

10-9H. The Logic Probe

The logic probe (*Figure 10-51*) is a basic tool for quickly testing most digital circuits encountered in RF work. A logic probe consists of HI and LO LEDs, which indicate when a signal is a 1 (HI) or a 0 (LO). A *TTL/CMOS* switch is included so that the probe will recognize which input voltage levels are appropriate for the type of logic under test. When the *MEMORY/PULSE* switch is set to *PULSE* detection and the logic probe is receiving a 50/50 duty cycle pulse train, such as a clock input, then both HI and LOW LEDs will light with equal brightness and the PULSE light will blink. If duty cycle is lower, then the LO LED will be brighter than the HI LED and the PULSE light will blink. If the *MEMORY/PULSE* switch is set to *MEMORY*, then a single pulse, which might normally be missed, will light the PULSE light. This LED will remain lighted until reset by switching the *MEMORY/PULSE* switch to *PULSE*. If the LEDs are not lighted, then there is a bad level or open circuit present, or you could be testing tristatable bus lines that are in their third state, which is high impedance. In this case a lack of standard levels is normal. Clip the operating power to the probe by simply attaching any positive 5V DC to 18V DC to the red lead, with the black lead going to circuit ground. Read the logic probe specifications to confirm particular operating voltage span for the probe in use. When attaching DC supply power to the probe when testing CMOS circuits, connect to the CMOS power supply (Vcc) circuit to obtain accurate HI/LO indications. (*CMOS LOW* is indicated by a voltage lower than 30% of Vcc; a *HIGH* is greater than 70% of Vcc — most probe circuits use Vcc as the reference voltage.)

10-10. Soldering

Even though it is second nature to most technicians, it may be worthwhile to reiterate proper soldering practice. Both through-hole and SMD (*surface mount device*) soldering and desoldering techniques will be covered. While the former is, once learned, very easy, the latter is usually such a complex task that many technicians find it next to impossible to perform proper SMD soldering or desoldering — with, or without, special tools.

Caution:
1. Follow ESD (electrostatic discharge) procedures when handling static sensitive components — always ground the soldering iron tip when working with ESD sensitive components, even if Zener input protected.
2. Most modern soldering tips are plated with nickel and do not normally need to be filed, while the old copper tips rapidly become pitted and must be filed to maintain their shape.

10-10A. Through-Hole Soldering

When soldering through-hole PCB components, use a 25W soldering iron with a tinned chisel tip. Tin by melting a small amount of solder on the tip to aid in efficient heat conduction between the iron's tip and the component's lead. As shown in *Figure 10-52a*, warm up the trace pad on the PCB and, at the same time, the component lead. Now, as in *Figure 10-52b*, place the rosin flux solder against the opposite side of the pad and component lead. Let the pad and the component's lead melt the solder, not the direct tip of the iron. Complete melting should take no more than two seconds if the iron is at the correct temperature. The solder will then flow around the leads. Remove tip from board. After the solder joint cools it should have a shiny silver appearance. If it appears to be dull, a cold solder joint has formed and should be reheated to reflow the solder around the connection. Keep the iron's tip clean (by wiping the tip on a moist sponge frequently) and well-tinned for reliable soldering.

For desoldering of through-hole components, three options are available. The first is a manually operated spring-loaded desoldering vacuum pump. Heat the solder joint with a soldering iron, placing the pump's tip into the molten solder, and releasing the pump's trigger. This sucks up the molten solder into a reservoir (which should be safely emptied periodically). The desoldering bulb is a simple version of this concept. The second technique employs a desoldering wick. Place a wick over the joint to be desoldered and arrange the iron on top of the wick. As the solder melts it is "wicked" up into the finely woven copper braid. Wick comes in rolls — use one section, snip off the used part and discard. The third method uses an electric desoldering station that melts and vacuums up the solder, in a single semiautomatic step, through a hollow tip and into a filter that must be replaced occasionally.

10-10B. SMD Soldering

Various methods have been developed to solder and desolder surface mount devices for repair purposes. The following procedures are proven, and do not require special equipment (which is frequently unavailable and of questionable value), except for "J" leaded devices.

To remove two-terminal SMDs, such as capacitors, resistors, and diodes, from a PCB: Quickly heat both ends of the surface mount device by rapidly alternating from one end to the other (*Figure 10-53*), then *twist* and pull up on the SMD with tweezers after the end connections are molten. Clean off the excess solder on PCB with solder braid. (**Caution**: *SMDs are generally epoxied to the PCB.*) There are also accessory soldering tips available that will heat both ends simultaneously, facilitating removal.

Follow this procedure to replace two terminal SMDs on a PCB. Melt a dab of solder on one pad. Place the SMD over both PCB pads. (*Figure 10-54a.*) Push down on the SMD as you heat the solder dab, while not touching the end termination of the component with the iron (or the delicate device can be damaged). The end of the SMD will now sink through the melting solder dab to the copper pad. Remove heat to the connection after 3 seconds. (Any longer and component and/or board damage can occur.) Complete the operation by carefully soldering the other end by heating just the copper pad (and not the component's termination), and placing the solder against the pad (*Figure 10-54b*) — allowing the solder to melt and flow into and around the end termination contact of the SMD. When set, the solder fillet should not significantly extend beyond the top of the SMD and must have a shiny, smooth appearance — with no apparent visible thermal damage to the device (such as cracks in the body or delamination of the end termination contacts).

Integrated circuit SMDs must employ other techniques for desoldering and soldering due to their multiple leads. The removal and replacement of SMD ICs of the gull-wing leaded SOIC (*small outline integrated circuit*) and QFP (*quad flat pack*) types, as shown in *Figure 10-55*, are feasible without special equipment. "J" leaded ICs are considered too difficult to perform without special equipment.

To desolder a gull-winged SMD, grasp the number 1 IC lead with tweezers. Heat the IC's lead at the point where it contacts the PCB pad; twist up on the lead after the solder liquefies. Continue until all leads are desoldered. Twist and pull up on the IC's body to break the epoxy bead. Timing and practice, as well as patience, are vital — or damage to the PCB is a certainty.

To place a new SMD IC, clean any solder from the copper pads with a wick and iron. Remove any old epoxy and check the pads for delamination from the PCB. Place the SMD IC over the pads and solder one of its pins in place. (Never pick up SMDs with the hand — use tweezers or a special vacuum pencil, or the leads may become bent or the IGFETs inside the IC damaged.) Confirm proper placement of all pins over the pads. Solder the remaining pins to their pads. Check for solder bridges and any incomplete joining of IC lead to pad by inspecting all work under a magnifier.

The use of .015" or less *eutectic* (63-37 tin-to-lead) rosin core solder and a (25W) 1/16" or less diameter soldering iron tip will make your SMT (*surface mount technology*) soldering chores a little easier.

A promising new method for removing SMDs is the patented Chip Quick ™ SMD removal system. A special liquid flux is spread on the device to be removed. A unique low temperature desolder/alloy is melted onto the SMD's pins with a soldering iron, combining with the original solder. The SMD will now pull off the board without damage to the copper pads. Complete by cleaning up the area with a soldering iron and braid or other desoldering technique. Wipe off the PCB area with cleaner solvent. Install a new SMD chip as above.

10-11. Digital Troubleshooting

Digital electronics is now integrated into many aspects of RF technology: Digital display metering, automatic channel scan and memory, on-board clocks and timers, digital modulation methods, spread spectrum communications, digital multiplexing, cellular communication, pocket radio, digital signal processing (DSP), etc. A common use of digital circuits in modern communications radios employs dedicated microprocessors, or single-chip microcontrollers, to accurately and smoothly control transceiver and receiver tuning. These digital ICs, at the request of the radio operator, send data commands to the frequency synthesizer to change frequency, while driving the radio's display. The digital ICs are extremely reliable and almost never fail. Most are ROM or EPROM based with their instructions burned in, with RAM registers available for user programming (such as favorite frequencies available — at the touch of a button) and as temporary registers for internal program utilization.

What appears to be a microprocessor failure is usually elsewhere. To troubleshoot, first make sure the IC is getting the proper V_{cc} (within 0.25V) with no noise and very little ripple (ordinarily not more than 1/50 of the amplitude of V_{cc}). Confirm that ground is not floating due to a bad solder connection or broken trace. (There should be zero volts when referenced to system ground — if not, then the ground line is floating.) Check all clock inputs for proper frequency, amplitude, and signal cleanliness. Inspect for proper I/O levels while the unit is under operation. If input levels are bad, trace back for shorts or opens on the traces or components. When a pin or line should be HIGH but is found to be LOW, check for a short to ground. Make sure no pins are just "floating" — generally noise will be seen on the oscilloscope, or the pin will be in a high or intermediate voltage state when it should be low. Confirm that the microprocessor resets when V_{cc} is cycled off and on by checking the reset pulse at the proper pin. (Do not depend on the power switch since the microprocessor will still be receiving its supply voltage — disconnect the power cord for several minutes to discharge the charged backup capacitor.) If any output levels from the IC are not correct, then lift the bad pin after checking all I/O ports for proper operation. Now check the pin for proper levels. If levels are now good then examine the trace paths "exiting" from pulled pin for shorts, or for any shorted coupling caps, etc. Rotate the radio's tuning knob. The logic probe should begin pulsing at the proper output pin (serial) or pins (parallel) to the PLL chip programmable divider input and to the frequency display or its display driver. If pin levels are still wrong after lifting the lead, and all other pins have been checked for proper levels, then the microprocessor should be replaced.

A few words about *Tri-statable* (also referred to as *three-state*) devices, since a chip so equipped can appear to be floating or open and may create confusion during digital troubleshooting. Tri-statable switches are necessary for most modern microprocessors to access their support chips, such as ROM, RAM, and I/O ports on a common bus line. The Tri-statable switch allows only one chip to output its digital word on the bus at a time. If there were no Tri-state outputs on a chip, then the chip outputs would not be isolated from the common bus: As soon as a chip's output went high, it would be shorted out by a low, which is usually ground or zero volts, allowing large currents to flow — with no way of telling whether the output was a one or zero. Tri-statable switches, on the other hand, are present as part of the output/input of any supplementary memory and I/O chips of a microprocessor, as well as in the microprocessors themselves. Chips outfitted with Tri-state bus outputs can supply a normal high (5V for TTL) and low voltage (0V), as well as a high impedance state, which acts as an open to the common bus line. In other words, the microprocessor must tell its support chip that it wants to read or write to that chip, or its outputs are switched "off". If chosen, the output would be switched "on", allowing either a 1 or 0 bit to be sent in to or out of the support chip and on to the common bus lines. As soon as that bit (or an entire word — one bit per line) is sent, the Tri-state switch is turned off and another is turned on so that the chip can also communicate. Only an external RAM, ROM, or I/O chip that receives an active low from the microprocessor into its *chip select* input will open that chip's Tri-state lines, allowing that chip to speak with the microprocessor. This permits other devices to output their signal, at different time slices, onto the common bus lines without interfering with each other and nullifying the need for each output to have its own line. The Tri-state level will be floating and is usually between 0.8 and 2.4 volts on an oscilloscope or DMM, though bad levels will be indicated on a *logic probe*'s LEDs. Some dedicated processors are single-chip types, with the ROM, RAM, and I/O all on one chip, and may or may not have Tri-statable output/input considerations. However, output enabling control pins, as well as address and data pins for supplementary memory and I/O chips, may be present — but most will go unused.

10-12. Electromagnetic Interference

Considering that electromagnetic interference (EMI) can sometimes simulate equipment failure, either continuous or intermittent, the technician should be familiar with its identification and cure. As well, the technician may be required to track down the interfering device and, hopefully, eliminate or at least attenuate the problem, either at its source or at the receiver. The technician must also be able to identify and cure any EMI that may be exiting any transmitters in the form of RFI (*radio frequency interference*).

Electromagnetic interference is generated by all forms of devices, and can adversely effect AM, SSB, CW and FM receivers, television sets, biomedical apparatus and many other types of equipment. EMI can cause anything from white noise in the receiver to a total annihilation of the desired input signal. Sources can be spurious generation, or even proper and clean fundamental transmissions, from any type of RF transmitter. EMI can also be generated from the operation of computers, light dimmers (thyristor circuits), fluorescent and neon lights, arcing in defective transformers or internal combustion ignition systems, and in all brush-type electric motors, to name just a few. In fact, anything that produces electrical sparking or fast current steps from zero to some value (or from some value to zero), will generate electromagnetic radiation up to very high frequencies. This radiation can be transmitted through the air or through the AC power line for extreme distances.

If the transmitter under test has been accused of creating RFI, then examine the transmitter's output directly using a spectrum analyzer. (See *Spectrum Analyzer* and *Attenuators*.) If all right, then attach a sniffer probe (*Figure 10-56*) or antenna to the analyzer, checking around the entire transmitting installation, for any spurious outputs. If found, confirm that the transmitter is well grounded and tuned, making sure that ground straps are not excessively long and not radiating RF, and that the transmitter cabinet is not radiating harmonics through open inspection doors (or even empty screw holes). As well, check that the antenna and its feedline are undamaged, and that all connectors are tight and clean. Establish that the transmitter is not generating parasitics or feedback oscillations. Check for master oscillator stability during modulation. Confirm that any low-pass filters, isolators or wavetraps used for harmonic or nearby transmitter frequency (transmitter-generated IMD) suppression, if present, are functioning as designed. Sometimes brute-force line filters will have to be employed to filter out any RFI from entering or leaving the transmitter through any power cables.

If the on-site transmitter is sending a clean signal, yet TVI (*television interference*) and RFI complaints continue, the complainant's radio receiver or television itself may not be properly shielded or filtered against fundamental overload or low level harmonics for use in an EMI filled world. If the transmitter is in the shop, it can be seen that a complete check of this nature, as outlined above, is impossible.

If the technician is on-site and suspects EMI instead of equipment failure while troubleshooting a receiver, and the receiver is blanking out intermittently, or there is excessive noise generated at its output such as a popping, whining, cracking, buzzing, or pulsing sounds, etc., then investigate for nearby noise producing devices. As mentioned above, check for light dimmers, brush-type electric motors of all kinds, arc-welders, adjacent clean high-powered transmitters (which can create fundamental overload and powerful, but legal, harmonic amplitudes) or low-powered but spurious generating transmitters.

Also look for passing noise-generating vehicles (standard and electronic ignitions creating electrical arcs); nearby computers (especially older types) or microcontrollers (standard in many kinds of electrical and electronic consumer equipment); starter-type fluorescent lights; high voltage neon signs (the longer the run of neon, the higher the possible voltage); and any device that uses an oscillator or frequency synthesizer, such as poorly shielded or defective television sets and radio scanners. The problem can be abated by utilizing a line filter at the receiver under test, or filtering its input with a high-pass or low-pass filter to attenuate any lower or higher frequency interference, or by trying out a bandstop filter to notch out the offending frequency. Confirm that the interference is not entering any of the receiver's cabinet openings and that the cabinet, if so designed, is properly grounded, with no ground loops present. (Some devices were not intended to have a grounded cabinet — check for "hot-chassis" transformerless or autotransformer designs.) Check that the feed cable from the antenna is undamaged. Use 0.01 µF ceramic-disk capacitors from the receiver's speaker to ground, and/or place ferrite beads around the speaker leads, or wrap any connecting cables around a ferrite rod or toroid (to attenuate any interference picked up by these cables). If necessary, replace unshielded cable with shielded type. Of course, if it is expedient and possible, attempt to neutralize any EMI generating sources at the point of origination by requesting the device's owner repair the defective equipment, such as arcing power line transformers or faulty transmitters. Or, if the offending device is on-site with the receiver and your company so authorizes, nullify the aberrant unit's EMI output: Use bypass capacitors on electric motors by adding 0.01 µF ceramic-disk capacitors from each commutator brush to ground or, if the brushes are worn and the commutator is dirty, then clean and repair. For neon signs, place 10 kΩ resistors in series with each supply lead. With fluorescent lights, confirm that the ballast and light tube are operating properly — if not, replace ballast or tube. (Older units employ starters and must be outfitted with bypass capacitors.) Replace or bypass offending light dimmers. For EMI-producing computers, place split ferrite toroids around all cables exiting the central processing cabinet of a computer (as close to the cabinet as possible). When feasible, utilize AC line filters on the offending equipment to prevent EMI from entering the common power line. If unfamiliar with the electrical equipment above, advise owner to have offending devices repaired or replaced by a competent electrician.

If receiving complaints that a new or old mobile receiver or transceiver installation is acquiring interference, confirm that the power leads to the units are well shielded or filtered, that the suppressor ignition wires and spark plugs are installed and undamaged, or that any after-market suppressor kits are installed and functioning. Make sure that all shields for the spark plugs, wiring, and other ignition devices, if so installed, are present. Establish that ground-strap bonding points (for maintaining proper ground) are still functioning. Verify that the alternator is not creating a "whine". Examine the alternator's connection integrity to the battery, while also confirming that any EMI filter present in its power lead is operating. If the on-board vehicle microprocessor or microcontroller is suspected of creating EMI, check with the vehicle manufacturer. (Transmitting from a modern vehicle may also cause adverse and unexpected results to any unshielded microprocessor controlled circuit, such as the ABS breaking and ignition systems.)

10-13. Important Troubleshooting Notes

The following notes are important to remember when troubleshooting RF circuits.

•*Aluminum* invisibly oxidizes quickly in contact with air, forming insulation. Always use toothbrushes and/or sandpaper at connection points to the aluminum chassis — both during new connection or a reconnection. As well, anodized aluminum chassis do not conduct electricity, as the anodization process creates an extremely thin layer of insulation. This must also be scraped off to supply a proper electrical connection to the chassis.

•Use only *nonconductive tuning tools* for adjustable capacitors and inductors — the added reactance will shift the proper tuning point of the circuit during alignment.

•*Transformers* employ color codes for their leads, but a replacement part may use a different color code.

•The same type of *transistor* may have a different lead arrangement than that replaced.

•Some *metal film resistors* are flame retarding and when changed out must be of the same type.

•*Wire wound resistors* have a lot of inductive reactance (unless wound non-inductively), and cannot normally be used at RF.

•In a large *rotor-stator air-tuned capacitor* there is a trimmer capacitor for each stage of the capacitor. These are used to adjust the minimum capacitance when the plates are completely unmeshed, and must be properly adjusted or tuning calibration will be inaccurate.

•*Mylar caps* have replaced paper caps and due to Mylar's relatively high absorption losses (energy lost in the dielectric) at higher frequencies, as well as inductive effects caused by their typical foil-wound construction, they must be used for lower frequency applications only. Non-inductively wound Mylar and paper capacitors are also available.

•*Ceramic-chip capacitors* have the highest operating upper frequency range, with ceramic disk capacitors just under the chip type. Mica and ceramic capacitors are employed at RF due to their low inductance and low absorption losses. Ceramic capacitors may also have a *variable temperature coefficient* (an increase or decrease in capacitance depending on temperature). N### equals a negative temperature coefficient; P### equals a positive temperature coefficient; NPO equals no temperature change. The numerical values (represented here by #) given after the N or P designations are in parts-per-million per degree Celsius change in *capacitance* from the reference temperature of 25°C. Confirm that any replacement capacitor has the same temperature coefficient, as well as the same voltage rating. (Replace a capacitor with a rating too low and the dielectric can be punctured, causing continued circuit failure in the future.)

•*Electrolytic capacitors*, especially the aluminum type, should have close to their voltage rating placed across them to maintain their oxide film, and thus their capacitance. They must never have a large AC voltage placed across their terminals, unless there is a larger DC voltage to maintain the necessary polarity as marked on the component. (There are also non-polarized electrolytic capacitors but, due to their internal structure being composed of two opposite polarity electrolytics back to back as well as two diodes, their capacitance per package size is less than a standard electrolytic.) Use only fresh aluminum electrolytics as replacements to avoid any with dried-out or leaking electrolyte. *Tantalum electrolytics*, however, have virtually unlimited shelf life. Most electrolytics have poor tolerance, and a leakage current of around 0.2 mA per μF of capacitance. (This is usually not a consideration because of its common application in filtering, bypassing, and coupling at low frequencies.)

•If a fault is found in directly coupled two-stage *DC audio amplifiers*, keep in mind that if the first stage ceases conduction its collector voltage would now increase. The second stage, being connected to the collector of the first stage, would begin to saturate, with the collector current now increasing toward maximum, which may destroy the second stage transistor.

CHAPTER 11
Support Circuits

Any circuit that is not absolutely necessary in receiving or transmitting a signal of a basic receiver or transmitter can be seen as a support circuit. Many support circuits are so invaluable that a modern transmitter or receiver could not adequately function without them, such as speech processing, mixing, AGC, and power supply circuits. Other circuits, which are not discussed here, could be termed as "bells and whistles," and are strictly for added operator convenience and are not indispensable for reliable and accurate communications.

11-1. Speech Processing

Overmodulation in AM causes *splatter*. Splatter is created by the additional harmonics produced in the baseband signal, which further modulates the carrier, creating extra sideband components and affecting a much wider bandwidth than desired. (See *Overmodulation*.) This produces adjacent channel interference, as well as making the primary AM signal less understandable.

Speech compression limits the maximum amplitude of the AM or SSB audio, while increasing the low amplitude audio signals. This maximizes the average power output of the transmitter while limiting maximum modulation to 100% or under. Speech compression functions similarly to standard AGC, but is placed in the audio stages. (*Figure 11-1.*) FM (and some AM and SSB) transmitters may employ a *speech clipper*. (*Figure 11-2.*) The clipper slices off any voice signal that is above a set design amplitude. This clipping action creates harmonics, which are removed by a low-pass filter at the output of the audio section. The low-pass filter is also used to limit the maximum frequency excursions of the modulating waveform. Another type of compression is employed in SSB, AM, and some FM transmitters, and is placed in the intermediate and radio frequency stages. It is referred to as ALC (*automatic level control*) and is also basically a standard AGC circuit, but located in the transmitter section. Automatic level control manages the gain of the transmitter's IF by applying a voltage to the gate, grid, or base of an IF amplifier. This is accomplished by taking the final's plate or collector voltage and rectifying and filtering it into a DC control voltage, then sending it to the IF amplifiers. ALC will lower excessively high amplitude output levels, but amplify medium and low level signals normally — in other words, a form of compression.

Since the modulation frequency in AM, SSB, and FM transmitters governs the transmitted bandwidth, some way to limit the maximum baseband frequency must be used. Any one of a variety of audio low-pass filter designs can be placed in the audio section for this purpose.

Figure 11-1. *Speech compression for a transmitter.*

Figure 11-2. *An FM transmitter's speech clipper.*

Figure 11-3. *A single-ended FET mixer stage.*

Figure 11-4. *Dual-gate MOSFET mixer stage.*

11-2. Mixers

Mixers are devices utilized to produce a sum and difference frequency from an input of two unlike frequencies. This process is referred to as *frequency conversion*, and is employed in both transmitters and receivers to decrease or increase a sinewave or complex signal's frequency. If the sum is used, it is referred to as *up-conversion*; if the difference is used, it is called *down-conversion*. Many AM and SSB transmitters use mixers to convert up to a higher frequency for transmission. In superheterodyne receivers a mixer is sometimes referred to as the 1st detector, and is utilized to convert a received signal to a lower frequency, called the intermediate frequency, or the IF. The conversion process, called *heterodyning*, is adopted for receivers because lower fixed frequencies are much easier to efficiently amplify than higher varying frequencies. This increases sensitivity and selectivity to obtain a narrower and higher gain response in the receiver, since the IF amplifiers can be tuned, filtered and optimized for a single low frequency.

A receiver obtains its IF through this process: any received input signal *beats* (heterodynes) with the local oscillator frequency in the mixing element, which can be a diode, transistor, tube, or MOSFET run in the nonlinear portion of its bias (for active devices, usually class AB). However, many modern RF mixers prefer to employ single or dual gate MOSFETs as their nonlinear conversion elements, due to their low mixing noise and wide dynamic range. This beating commonly produces, at the mixer's output, the signal frequency, the local oscillator frequency, and the sum and difference frequencies of the signal frequency (with any attendant sidebands) and the local oscillator frequency. The new desired translated frequency, consisting of the carrier and any sidebands at the (generally) difference frequency, is filtered at the output of the mixer and sent through fixed tuned IF amplifiers. All other unwanted frequency components, such as the sum of the signal and LO frequencies, as well as the original signal and LO frequencies, are removed by the passband selectivity of the IF filters.

There are different classifications of mixers. The output of *unbalanced* mixers, such as that described above, are f_s, f_{LO}, f_s - f_{LO}, f_s + f_{LO}, with very little isolation between the mixer's input and output ports, resulting in signals which can radiate or interact with any other input or output port. A mixer that is *singly*-balanced significantly attenuates either the original signal *or* the LO, and passes the resulting mixing products on to its output. A *doubly*-balanced mixer (DBM) furnishes great IF-RF-LO port isolation while outputting *only* the sum and difference frequencies of the signal and LO — making the job of filtering a far simpler task.

Amplitude modulated and single-sideband transmitters cannot use simple frequency multiplication circuits as are employed in FM circuits (to increase the signal's frequency after modulation). AM and SSB must apply the heterodyning techniques described above — or major distortion of the delicate modulation envelope would result.

11-2A. Single-Ended FET Mixer

In the *single-ended FET* mixer of *Figure 11-3*, the RF input signal is dropped across the tuned input tank and coupled to the FET's gate, while the LO output is dropped across R_1 and into the source lead. The

Figure 11-5. *One type of single-ended transistor mixer stage.*

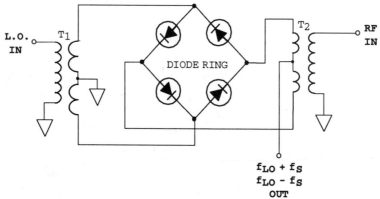

Figure 11-6. *A doubly-balanced mixer stage with diode ring.*

Figure 11-7. *An autodyne converter: A combined mixer and local oscillator.*

mixed signal is then taken from the drain and placed into the output tank, which is tuned to the desired IF frequency. All other mixing, signal, and LO frequencies are discarded by the tuned tank. The IF is now induced across the secondary of the output transformer and on to the IF amplifiers. *Dual-gate MOSFET* mixers, such as that shown in *Figure 11-4*, can also be used, with the LO output being injected into the first gate and the RF signal into the second gate. The difference frequency, equal to the IF, is removed from the secondary of the tuned output transformer of T_1 and sent on to the IF strip for further amplification and filtering.

11-2B. Single-Ended Bipolar Transistor Mixer

The *single-ended transistor* mixer of *Figure 11-5* injects both the signal and the LO into the shared base, where they are mixed by the nonlinear class AB biased transistor. All mixing products, as well as the original signal and the LO frequency, are in the output — but only the IF is passed to the next stage by virtue of the selectivity of the output tank circuits.

11-2C. Diode-Ring Mixer

The doubly-balanced mixer of *Figure 11-6* uses a *diode-ring* to perform frequency conversion of the input signal. Although diode mixers have a conversion loss (the output amplitude is less than the input amplitude), they have the advantage, as stated previously, of only outputting the sum and difference frequencies and attenuating the original carrier and LO frequencies, while offering IF-RF-LO port isolation (which is of primary importance in eliminating any LO frequency from being radiated from the antenna) — as well as a general increase in mixer stability. This type of mixer can be operated at VHF and above using *hot-carrier* or other high frequency diodes. Doubly-balanced mixers are available in a module, with the diodes and transformers already balanced in a single package.

Other diode mixers are available that use either a single diode, or two diodes and a center tapped transformer. However, they are not doubly-balanced, and must also be supplied with a very high amplitude LO output for proper mixing. They are, nevertheless, low in cost and need few components.

11-2D. Autodyne Converters

Autodyne converters are utilized in low cost receivers that operate below 30 MHz. They employ a single transistor for combined mixing and local oscillator chores. (*Figure 11-7*.) The LO frequency is selected by T_2 and C_4; the IF by the resonant tanks consisting of T_1, C_2, and C_3.

11-2E. On-Chip (IC) Converters

On-chip converters boast on-board voltage regulation, a local oscillator, and a mixer (*Figure 11-8*) — all in a single integrated circuit. By adding a few external discreet components the circuit is complete. Some types of IC mixers are usable up to and beyond 0.5 GHz. The LO frequency is set by an external crystal or an LC tank, and the IF is chosen by a ceramic, crystal or other method at the mixer's output.

Figure 11-8. *An IC with local oscillator, mixer, and voltage regulator.*

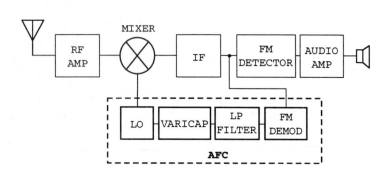

Figure 11-9. *An AFC circuit for an FM receiver.*

Figure 11-10. *Graph of base-bias voltage versus amplifier gain for AGC.*

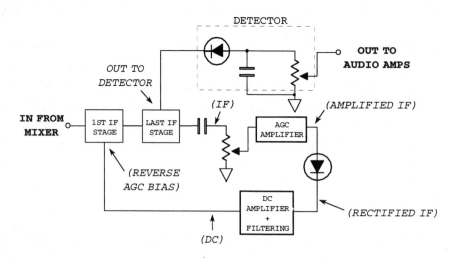

Figure 11-11. *A popular AGC circuit using the IF signal.*

11-3. Automatic Frequency Control

AFC (*automatic frequency control*) is adopted to maintain the local oscillator of older, or some newer but low cost, receivers on-frequency without employing frequency synthesis.

Any receiver that can be tuned over a wide range using LC oscillators needs some method of remaining on frequency, even with internal and external temperature changes (which would create frequency drift that is unacceptable in modern equipment). A superheterodyne receiver with local oscillator drift converts the incoming signal to an IF that is not exactly at the IF amplifier's center frequency — thus causing signal attenuation and distortion of the received signal.

The AFC of the FM receiver of *Figure 11-9* can detect the FM from the last stage of the IF section with a separate demodulator, or a slightly altered circuit configuration can tap into the receiver's own detector. Looking at *Figure 11-9*, the AFC's demodulator will vary its output voltage in step with any drifting of the intermediate frequency, which is caused by LO drift, and sent to a low-pass filter, which filters the output of the demodulator into a DC control voltage, adding-to or opposing the varicap's rest frequency bias. This varicap has been placed across the tuned circuit of the LO. The varicap, and its bias control voltage from the filter/demodulator, has a certain amount of control over the frequency of the LO. If the local oscillator drifts off frequency, the demodulator will vary from its resting output voltage and feed a correction voltage to the varicap, which then changes its capacitance, forcing the LO back on frequency. Most modern equipment uses the *phase-locked loop* (PLL), which is crystal steady and tunable over a broad range.

11-4. Automatic Gain Control

AGC, also referred to as AVC (*automatic volume control*), is utilized in virtually all modern receivers. Without AGC, strong signals would overdrive the receiver and cause distortion, while weak signals would be virtually undetectable by the listener. The dynamic range of receivers would be extremely low.

AGC functions due to a certain simple but important transistor characteristic: Increasing the transistor's collector current increases the transistor's gain. An easy way to increase the flow of current in the collector is to increase the forward bias at the base, since increasing base current increases collector current. A point is eventually reached where this ability levels off and then gain begins to decrease if collector current is increased any further. (*Figure 11-10.*) However, it must be emphasized that this base and collector current is direct current that increases the gain, and not the signal current. This current is produced by a *DC bias voltage* impressed at the base of the transistor by the AGC circuit (as shown in *Figure 11-10*). Indeed, in some AGC amplifiers the only DC bias, which fixes the Q-point, is set strictly by this AGC voltage.

As stated above, increasing collector current increases the transistor's gain, yet a point is reached where this ability tapers off. Gain then begins to decrease if collector current is increased further. Because of this ability to increase and decrease gain by increasing or decreasing collector current, it only follows

Figure 11-12. An AGC circuit using the detected audio signal.

Figure 11-13. An IF amplifier employing reverse AGC.

Figure 11-14. An RF amplifier with forward AGC.

Figure 11-15. Dual-gate MOSFET IF amplifier with its second gate controlling stage gain.

Figure 11-16. Simple single-diode IF AGC.

that there are two ways to apply AGC: *reverse* and *forward AGC*. Reverse AGC generally controls the gain of the IF section, while forward AGC commonly controls only RF amplifiers. Reverse AGC lowers gain by decreasing the bias on the base, which decreases collector current, and thus gain. Forward AGC increases forward bias on an amplifier, which increases collector current, and thus reduces stage gain. The AGC voltage can be tapped off of the last IF stage (*Figure 11-11*) or, in some receivers, after detection by the detector. (*Figure 11-12.*) In *Figure 11-11* the IF signal is tapped, amplified, rectified, filtered to obtain a DC voltage then, in some receivers, further amplified — and applied to the base of the 1st IF amplifier through an *AGC bias line*. (Actually just a wire or trace on a PCB board.) In many cases AGC is also administered to the second IF stage as well. When the AGC is taken after the detector as in *Figure 11-12* (a technique frequently used in AM radios), a separate AGC rectifier is not needed — operation is basically the same as above. FM receivers frequently tap the signal amplitude for the AGC circuits from the main discriminator. In *Figure 11-13* the IF Class A amplifier shown utilizes reverse (negative, for NPN transistors) AGC to subtract from the base's normal positive bias. As a strong signal is received, the rectified and filtered negative (DC) AGC voltage is applied to the base of the transistor. The stronger the received signal the more negative the AGC voltage. This voltage opposes the positive bias voltage at the base (which bias's the transistor at its Q point, allowing a set amount of collector current to flow). As the base becomes less positive, less DC collector current flows, which lowers the transistor's gain. When AGC is applied to an NPN transistor, typical measured values at the base will be +1 for minimum gain and +3 volts for maximum gain.

For forward AGC, the positive (for NPN transistors) DC bias is increased at the base as a strong signal is received, which increases base current, which increases collector current, which decreases gain. The RF transistor amplifier of *Figure 11-14* utilizes this type of automatic gain control.

Entire IF sections are now obtainable on a single integrated circuit. Most of these IF ICs have pins available for manually setting the AGC level with an external potentiometer. This will adjust the amount of AGC amplification desired (very little AGC for distant stations — a lot for strong local stations). As well, regular AGC will still lower gain even when weak input signals are received — *delayed AGC* corrects this common problem by not supplying any gain reduction until a set signal amplitude is reached. This assures that low level signals will receive maximum amplification. The threshold input signal voltage needed to activate AGC is adjustable by attaching a resistor or trimmer to the proper IC pin. Accessory pins may also be available for feeding AGC generated voltages into *tuning indicators*. Considering that the greater the input signal amplitude the greater the generated AGC voltage, these indicators can consist of a simple LED that becomes brighter when a maximum signal strength is reached. The AGC output voltage may also be fed to a voltage or current meter, referred to as a *signal strength indicator*, and suitably calibrated in *relative signal strength units*, to show to the operator the intensity of the received signal.

Dual-gate MOSFETs are also commonly used in AGC controlled IF amplifiers. (*Figure 11-15.*) The MOSFET's second gate can be fed a positive AGC voltage, which increases stage gain, or a negative voltage, which decreases gain.

Figure 11-17. A squelch circuit using AGC for signal detection.

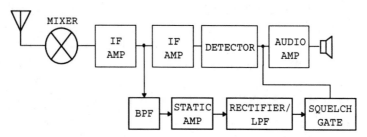

Figure 11-18. A noise-controlled squelch circuit.

Figure 11-19. Adjustable squelch circuit utilizing an IF IC's signal level output and a comparator.

Another form of AGC (*Figure 11-16*) exploits a signal diode. When a low signal is received, CR_1 is off and offers a high resistance to ground. However, when a large enough signal is received, CR_1 is biased "on", acting as a shunt resistance to ground for the greater IF signal.

See also *Section 6-2* for *SSB AGC* on a description of popular single-sideband automatic gain control schemes.

11-5. Squelch

Squelch, or muting, circuits are found in many AM and FM communication receivers. Squelch "removes" the annoying static that occurs when the radio is not receiving a signal, since gain is then normally at its highest. (See *AGC*.) In a common squelch circuit, the audio amplifiers are disabled in some way to remove any noise from reaching the speaker. Certain audio amplifiers are disabled by applying a reverse bias to the base of the stage, others completely cut off the supply voltage to these audio sections, while still others simply prevent the noise energy from ever reaching the audio amplifiers by either shorting the static to ground, or blocking it with a series pass transistor. Most exploit the DC output of the AGC circuits to detect whether or not there is any RF input signal present. In the squelch circuit of *Figure 11-17*, AGC operation occurs as follows: When there is a low or absent input signal level, the negative AGC bias is decreased, and is almost nonexistent (allowing the IF stages to be biased at maximum amplification). This AGC level is wired into the AGC IN of the squelch amplifier and on to the *squelch gate*, which is basically a transistor switch and is normally biased ON when low input signals are present. The low negative AGC level is not high enough to turn OFF this squelch gate, so the noise signal is shorted to ground through the low resistance of the ON transistor before it can even reach the audio amplifier stage. Thus, no audio signal reaches the speaker. However, when an RF signal of the proper amplitude is received, the AGC now generates enough voltage to cut off the squelch gate, allowing the detected audio to pass into the audio amplifiers without being shorted to ground. Other squelch circuits use a bandpass filter tapped into the IF section (*Figure 11-18*) to pass the high frequency noise (above 6 kHz) into a static (noise) amplifier, and on to be rectified and filtered to a DC control voltage. The control voltage is applied to a squelch gate which, if no signal is being received (always a high noise condition), is then biased into conduction, shorting the detector output to ground. This removes the audio signal from the input to the first audio amplifier, thus quieting the audio output. Any desired (or undesired) audio modulation received would instantly attenuate the high frequency noise by replacing it with the 3 kHz or lower audio frequencies, which would shut off the squelch gate, allowing the audio to reach the audio stages and be amplified. Others methods employ IF integrated circuits. (*Figure 11-19*.) The IF IC outputs a voltage level from one of its pins that corresponds to the received signal strength and into a comparator, which is adjustable for the preferred squelch threshold by the level control R_1. If Input 1 is below the reference level as set by the voltage divider of potentiometer R_1 at Input 2, which sets the level of squelch desired, then the comparator swings close to its positive rail (+Vcc), switching ON the squelch gate, shorting the collector of Q_1, the voltage audio amplifier, to ground through diode D_1. This removes the audio signal from the input to the next stage, the power amplifier, which quiets the speaker. Other more advanced IF ICs have built-in squelch circuits (and a first audio amplifier) that simply requires an external potentiometer attached to the appropriate pin to adjust squelch level. These are just a few of the many circuits developed to mute a receiver.

Figure 11-20. *An entire regulated power supply.*

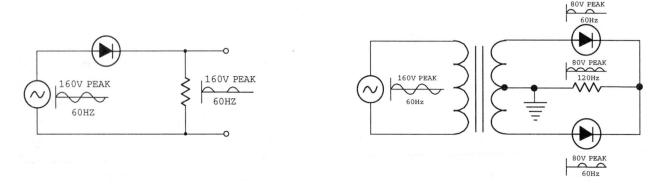

Figure 11-21. *A half-wave rectifier.*

Figure 11-22. *A full-wave rectifier.*

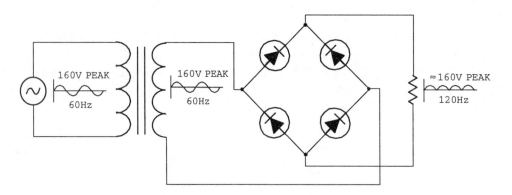

Figure 11-23. *A bridge rectifier.*

11-6. Power Supplies

Since batteries can only supply current for a limited period of time, the majority of RF equipment runs off of the AC mains through power supplies. Most power supplies convert this AC mains voltage into the desired DC levels needed to furnish power to the electronic circuits at a constant and regulated voltage. Banks of lead-acid batteries, however, are utilized extensively for backup power when a failure in power supply delivery occurs.

The average power supply consists of a transformer (for voltage conversion), a rectifier, a filter, a regulator, and ordinarily some form of protection circuits to safeguard both the PS and its load. A complete power supply is shown in *Figure 11-20*. It contains the transformer needed to convert the AC mains voltage of $120V_{RMS}$ into the desired voltage. The bridge rectifier circuit then converts the AC into a pulsating DC, which is then filtered by the filter network into a steady DC. The three-terminal regulator holds the output voltage steady, while C_3 suppresses any output oscillations, as well as aiding in regulation. RB, the bleeder resistor, is used to draw a fixed current, helping to stabilize the output voltage while draining dangerous voltage levels from the filter capacitors. Each of these circuits will be examined in more detail below.

11-6A. Rectification

There are three common ways single-phase (household) AC power can be converted into pulsating DC power, which is the first step in the power supply chain of obtaining a smooth DC output voltage, and is referred to as rectification. *Half-wave* rectification (*Figure 11-21*) supplies a difficult to filter 60 Hz output, but at a peak voltage that is equivalent to the incoming AC peak voltage from the power line or transformer, while requiring few components (a single diode). *Full-wave* rectification (*Figure 11-22*) delivers an easy to filter 120 Hz output, but at only half the AC voltage peak of the input (because of the transformer's center tap), while utilizing an expensive center-tapped transformer. Bridge rectification (*Figure 11-23*) gives an easy to filter 120 Hz at the full input AC peak voltage levels, and is by far the most common rectifier in modern, quality equipment.

11-6B. Filtering

Because pulsating DC power is unacceptable for most electronic circuits, filtering is needed to remove this pulsating component while supplying a constant unvarying current. A basic power supply filter is shown in *Figure 11-24*. This filter will remove any AC component from the rectified output, while permitting the DC. The filter consists of C_1, which filters most of the ripple, and R_1 and C_2, acting as an AC voltage divider: any ripple remaining after C_1 is dropped across R_1 (due to its higher AC resistance over that of C_2's lower reactance to the ripple frequency). Replacing R_1 with an inductor would have basically the same effect. However, RC filters work best with a low current drain, while LC filters will continue filtering even when the current drain is at a high level. Placing the capacitor at the input, as in the above example, will supply a higher final DC output, but with inferior voltage regulation. Placing an inductor at the input will supply a lower output voltage, but will furnish better voltage regulation.

Figure 11-24. *A common power supply filter.*

Figure 11-25. *A rectifier/filter that furnishes both negative and positive voltages.*

Figure 11-26. *A basic and low cost Zener shunt regulator.*

Figure 11-27. *A series-pass transistor regulator.*

A rectifier-filter variation that can be used for devices (such as op-amps) that require both negative and positive power supply voltages (*Figure 11-25*) simply adds a second half-wave rectifier. This second half-wave rectifier passes the negative portion of the AC wave, and filters it through C_1.

11-6C. Regulation

Most modern electronic circuits will only function reliably if they are provided with a supply voltage at a constant and unvarying amplitude. Considering that an unregulated PS's output voltage will change when either the AC line voltage fluctuates or the resistance of the load varies, regulators are needed to prevent, or at least lessen, these changes. Regulators also assist the filters in smoothing the output voltage.

There are many different types of voltage regulators. The simple *Zener shunt voltage regulator* of *Figure 11-26* drops excess voltage across the resistor RS as current increases through both the Zener and R_S. This occurs because the sum of the Zener voltage and the voltage drop across R_S must equal the input voltage ($V_{Rs} + V_Z = V_{IN}$). The Zener changes resistance as current varies to keep its Zener voltage (VZ) constant which, considering it is parallel with the load, keeps the load voltage constant. In other words, as Zener current increases, Zener resistance decreases — if Zener current decreases, then Zener resistance increases, thus keeping the voltage drop across the Zener (and the load) constant. ($V = IR$.) VIN must be slightly higher than V_Z for this Zener regulator to stay in regulation.

With the *series-pass transistor regulator* of *Figure 11-27*, if the voltage across the load should increase for any reason, the regulator's output voltage will remain relatively steady. Since the voltage across the load must equal the voltage drop across the series-pass transistor's (Q_1) E-B junction, subtracted from the voltage drop across the Zener (V_Z) — and this V_Z value cannot change — and considering that the base voltage is set by this Zener, then any increase in the regulator's output voltage will make the emitter that much more positive than the base. This is the same as making the base less positive, resulting in less B-E voltage, producing less emitter current and a subsequently larger voltage drop across the series pass transistor. This means a decrease in any attempted increase in voltage across the load.

The *series-pass regulator* with *feedback* block diagram (*Figure 11-28*) shows the major sections to this type of popular regulator. The circuit itself (*Figure 11-29*) consists of R_1, R_2, and R_3 forming a sample and adjust (voltage divider) circuit to set the desired output voltage. By sliding the wiper of the adjustable (pot) R_2 downward, the voltage on the base of Q_2 decreases, reducing its forward bias so that Q_2 will conduct less, causing its collector voltage, and the base voltage of Q_1, to increase. With the bias of Q_1 now increased and with Q_1 subsequently conducting harder, more current is sent through the load, increasing V_{OUT}. The Zener sets the reference voltage at the base of Q_2 by keeping the transistor's emitter clamped at this voltage, while R_5 fixes the idling current through the Zener. R_4 is both the collector resistor for Q_2 and the bias resistor for the base of Q_1.

Series-pass regulators with feedback are a large improvement over the simple Zener shunt regulator, as very little current is now wasted through the Zener shunt path. Series-pass regulators with feedback also allow, unlike both the Zener shunt and the basic series-pass regulator, a simple scheme to obtain a

Figure 11-28. *A block diagram of a series-pass regulator with feedback.*

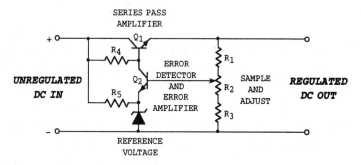

Figure 11-29. *The complete circuit of a series-pass regulator with feedback.*

Figure 11-30. *The popular three terminal regulator with adjustable voltage output.*

Figure 11-31. *A 723 IC voltage regulator.*

variable output voltage than can be changed anywhere from a little above the Zener's V_z to just under the unregulated supply voltage. These feedback regulators also furnish improved regulation due to the error detector and its amplifier, since any change in voltage is amplified and fed into the base of the series-pass transistor.

Three terminal regulators (or *3Ts, Figure 11-30*) are the most prevalent type of regulator because of low cost and simplicity. For any circuit needing a supply that must deliver three amperes or less (or higher currents with other external components), the 3Ts are the regulators of choice. These IC regulators contain internal current limiting and thermal protection circuits (if internal power dissipation becomes to high, then the regulator automatically shuts-down), and are lighter in weight and smaller in size, as well as much more efficient, than discrete regulators.

In the 3T circuit of *Figure 11-30*, R_1 and R_2 set the fixed output voltage, while R_2 can be varied to adjust this output voltage over a certain range. Both positive and negative voltage output versions, with common values of 5, 8, 12, 15, and 24 volts, are available.

Another widespread type of IC regulator is the multipurpose *723 adjustable voltage regulator* of *Figure 11-31*. Operated alone it can supply up to 150 mA to a load, but with a supplementary series-pass transistor almost any desired load current can be accepted. With various external passive components, it is commonly utilized as a 2 to 37 volt series linear regulator; or as a current, shunt, or even switching regulator. The 723 contains on-board current limiting set by an external resistor.

Highly efficient *switching power supply regulators*, some of which are 1/3 the weight and size of certain linear regulators, are found in an increasing number of equipment. They perform regulation by switching their output completely on and completely off while varying the resulting pulse's duty cycle or the number of pulses per second in step with the output load requirements. Most operate with switching speeds between 5 kHz and 500 kHz. Due to the switching noise that can be produced, their use is still limited in certain communication gear. Switching power supplies demand serious shielding, filtering of all input and output leads, and short internal leads, to minimize RFI generation into surrounding space. *Figure 11-32* is one type of switching regulator. Q_1 is the switching pass transistor. D_1 protects Q_1 from the inductive kickback of L_1. (Opening any inductive circuit causes a current pulse, producing damaging arcing in unprotected circuits due to $t = L/R$; when a charged RL circuit is opened, in this case by the switching transistor, the resistance is infinite, causing an extremely fast discharge time for the coil, giving off a very high spike of voltage.) L_1 and C_1 filter the pulses coming out of Q_1 to DC, while the voltage divider of R_1 and R_2 program the desired output voltage. The regulator functions as follows: when the load voltage decreases, the load decreases in resistance or the input voltage to the regulator declines. The comparator senses this change in output as a drop in voltage across R_2 below that of the Zener reference voltage, which then switches on the VCO, starting it oscillating and outputting pulses into Q_1. Q_1 then pulses on and off, outputting pulses into the filter network consisting of L_1 and C_1, which filters the pulses into a DC level. By increasing the pulse's duty cycle into the LC filter, the average DC output voltage increases. Much of this circuit, as well as similar switching schemes, can be obtained in IC form.

Figure 11-32. *An example of a switching regulator circuit.*

Figure 11-33. *A popular switching regulator IC with bias components.*

Figure 11-34. *A discrete regulator equipped with current limiting and overvoltage protection.*

IC switching regulators allow the substantial reduction in the size of any regulator heatsinks needed over that of a similar linear regulator IC in a power supply, and typically can furnish over 80% efficiency levels at around 3A output currents (in other words, little power is wasted as internal heat), with very wide input voltage ranges of between 4 and 40V possible. Most include built-in current limiting and thermal protection. *Figure 11-33* shows a common IC version of the discrete circuit of *Figure 11-32*. Such a regulator would be found in many switching power supplies, along with its characteristic support components arrangement. As discussed above, R_1 and R_2 program the output voltage value, with the feedback line between R_1 and R_2 supplying feedback to the regulator's internal comparator circuits. (R_2 can be made adjustable to vary V_{OUT}.) L_1 and C_{OUT} filter the output, while D_1 protects the IC from the inductive kickback of the discharging L_1, furnishing a return path for the load current when the IC switches off. C_1 is needed for regulator stability at higher current draws. If undesirable ripple amplitudes are still present at the output of this circuit, then a supplementary LC ripple filter is sometimes added after R_1 and R_2. With this breed of regulators, it must be understood that when an output capacitor is found to be defective, it *must* be replaced with the exact same type, or higher ripple outputs may result, as well as possible regulator instability due to variations in the capacitor's *equivalent series resistance* (ESR). The diode, D_1, must also be replaced with the same type, or instability and RFI may occur. This is caused by diode switching speed variations from diode to diode.

11-6D. Circuit Protection

If either the load or the power supply should have a failure, high currents could flow into the opposite device, destroying it before any fuse could open. There are two common circuits adopted for load and power supply protection.

Adding a *current limiting* circuit (*Figure 11-34*) to the above discreet regulators results in protection for the power supply if the load should develop a short. Without current protection the power supply's series-pass transistor could be destroyed, even before the fuse opened, if any low resistance path formed in the load. If current increases to the point where the voltage drop across R_6 is above 0.6V, the base-emitter junction is turned on, causing current to flow through R_4, making the base of Q_1 less positive, thus decreasing the collector current of Q_1. This action will lower the current within the power supply to a safe level.

Overvoltage protection is used to safeguard the load if the power supply should malfunction. If the power supply voltage rises above its design value, which in *Figure 11-34* is above 6.2V, the 6.2V Zener will begin to conduct, causing a voltage drop across R_7, which latches the SCR on, shorting out the load from the power supply circuit through the SCR, causing the PS fuse to blow. It is often called a *crowbar* circuit because it completely shorts out the power supply.

11-7. Digital Signal Processing

In *digital signal processing* (DSP) the desired received signal is enhanced by applying digital technology. Signal-to-noise and fidelity are improved, as is adjacent channel interference. The audio signal is placed into the DSP circuits and *sampled* (converted to digital pulses by an A/D converter). The

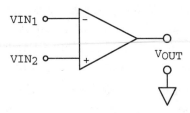

Figure 11-35. An operational amplifier configured as a comparator.

Figure 11-36. A passive diode clipper.

Figure 11-37. An active transistor clipper.

Figure 11-38. A Zener slicer circuit.

Figure 11-39. A diode clamper circuit.

Figure 11-40. An AC-to-DC voltage-doubler circuit.

Figure 11-41. A simple discrete constant-current source.

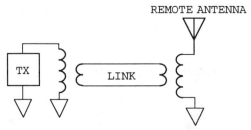

Figure 11-42. Link coupling between a transmitter and its antenna.

onboard microprocessor then processes these signals using algorithms (a series of steps employed to solve a problem) to remove any noise, and most of the interference. The processed output is then fed into a D/A converter to change the digital data of the converted audio signal back into an analog audio frequency. This recovered audio is then amplified and outputted to a speaker or headphones. DSP is capable of processing both incoming voice and digital signals, as well as video, with built-in DSP chips placed within the receiver, or through add-on external units.

11-8. Miscellaneous Circuits

There are many other important circuits that contribute to the operation of a transmitter, receiver, or transceiver. These miscellaneous circuits are encountered in most communications gear, and a quick explanation of their operation is appropriate to aid in further understanding and troubleshooting radio equipment.

11-8A. Comparator Amplifier

Comparator amplifiers (*Figure 11-35*) exploit an op-amp in their open loop (no feedback) configuration to compare two input voltage levels. Due to an op-amp's very high gain without negative feedback, *any* difference in potential between its inverting and non-inverting input causes saturation — either to its maximum negative or maximum positive supply voltage, depending on which input goes high. The comparator can be employed as a *zero-crossing detector*: when the input signal goes even slightly positive, the output voltage will switch to its full positive rail (which is near Vcc), and visa-versa for a negative crossing. The comparator is also used as a *level detector*: a reference voltage is placed at one input, while the voltage to be sample is placed at the other. Whichever input contains the higher amplitude will saturate the output at that corresponding polarity.

11-8B. Clippers, Slicers and Clampers

A *clipper* clips off one or both alternations, and is used to set a voltage to be no higher than to some design value. *Figure 11-36* shows a *diode clipper*, which only allows one alternation to pass while clipping of the top of the other alternation at 0.7V (if a silicon diode is utilized). In *Figure 11-37* an *active* clipper, which is basically an overdriven amplifier, will cut off the top of both peaks while placing the AC wave on a DC level. (The clipped wave never crosses zero volts.) However, only one peak may be clipped, depending on the bias applied to the transistor (to create either saturation or cutoff).

A *slicer* (*Figure 11-38*) slices off both AC alternations by using two Zener diodes consisting of any desired Zener voltage. The first alternation is passing through the reverse-biased first Zener, dropping 15V, while then passing through the second Zener in the forward direction, which drops another 0.7V. Since the two diodes are in series, and any load present is in parallel, a total of 15.7V is the maximum amplitude that is allowed to be passed from input to output in either the positive or negative alternations of the AC wave.

Figure 11-43. A transistor RF switch.

Figure 11-44. A transistor switch for DC.

Figure 11-45. A diode switch for RF.

A *clamper* (or *DC restorer, Figure 11-39*) clamps the incoming AC to some positive or negative DC level. It converts an AC wave into a varying DC wave by *clamping* the output to zero volts or below (or above, if the diode is reversed). First, the capacitor is charged up by the input signal as it flows through the diode. Then, when the input signal drops and can no longer forward bias the diode, the capacitor discharges through the resistor, creating a DC level for the input signal at the output load. The capacitor must be of a relatively high value, or distortion of the output wave will occur.

Occasionally a clipper is considered to be a circuit that limits one alternation above a certain set amplitude, while a slicer is considered to be a clipper that limits both alternations above a certain set amplitude. Slicers and clippers are commonly used as *limiters* in audio and FM IF circuits, or to reshape certain waveforms.

11-8C. Voltage Multipliers

Voltage-doublers (Figure 11-40) are utilized when increased voltages are needed in low current circuits. A peak AC voltage at the input is rectified and filtered into a *2 x V_{PEAK}* DC voltage at the output. The voltage-doubler works as follows: an incoming AC voltage forward biases the first diode, allowing current flow to charge the first capacitor. The second diode is reverse biased and cannot conduct. When the polarity of the AC reverses, the first diode becomes reverse biased and does not conduct, while the second diode is now forward biased and passes current, charging up the second capacitor to the opposite polarity to that of the first capacitor. The voltage is now taken from between the two oppositely charged capacitors. This output voltage is the difference between the two capacitors' potential difference. Doubling of the input voltage occurs, as well as the filtering of the AC into a DC, caused by the ability of the capacitors to store a charge. (If the capacitors are large enough to supply the load with a relatively steady current.)

Multipliers, in different configurations, are commonly available up to voltage quintuplers.

11-8D. Constant-Current Source

Certain circuits and test equipment need a constant-current source to function properly. The discrete constant-current source of *Figure 11-41* functions because there is a consistent 5.6V applied to the base, supplied by the reverse biased Zener. Consequently, a constant five volts is dropped across the emitter resistor and a constant 0.6V is also dropped across the base-emitter junction of the transistor ($5.6V_Z$ - $0.6V_{BE}$ = 5V). As a result, the collector current will be a steady 5 mA ($I_C = 5V_E/1K\Omega = 5mA$) through the load, no matter how much the load resistance may vary (up to a point, of course).

There are many other configurations to implement constant-current source circuits and current regulation, most of which are available in IC form or as an integral part of numerous other ICs (such as differential amplifiers).

11-8E. Link Coupling

Link coupling (*Figure 11-42*) can be placed between a transmitter and a remotely located antenna to transfer energy over long distances to the antenna. Link coupling consists of a few turns of wire connected, commonly, to a coaxial cable. This will minimize losses due to inductive reactances in long RF cable runs. Link coupling will lower harmonic transmission to the antenna, as the links have a low value of capacitive coupling between the coil output of the transmitter and the small turns of the link coils.

Link coupling is also adopted to couple test equipment, such as oscilloscopes and spectrum analyzers, to a transmitter — with almost no loading on the circuit under observation.

11-8F. Electronic Switches

Not only does the transistor perform as an amplifier of AC and DC signals, but it is also capable of switching DC, AC, and RF signals. *Figure 11-43* presents a simple circuit for switching RF currents. When the transistor's base is grounded through the manually operated mechanical switch, the NPN transistor does not conduct and does not allow any RF to pass to its output. However, when the mechanical switch is placed across the positive supply, base-current instantly begins to flow, and the transistor quickly saturates, creating a very low emitter-to-collector resistance. This permits the RF signal to pass to the output. Transistor switches can be used to remotely switch RF currents when long wire or trace runs containing RF is undesirable.

DC, especially for amplifier supply voltages, may also be switched. The transistor switch of *Figure 11-44* is activated when SW_1 closes, supplying a positive bias voltage to the base, turning on the transistor and allowing DC current to pass through the load.

Instead of directly placing a positive voltage on the transistor's base with a mechanical switch, a positive voltage can be sent to the base in response to other circuit conditions, such as when a transceiver changes from receive to transmit — the transmitter section now needs power, and not the receiver. The transistor switch can supply this power, as can the *diode switch*.

A very important function of diodes in RF is likewise for switching applications. Diodes are utilized to switch-in different values of crystals, circuits, or entire sections, into a signal path. *Figure 11-45* shows such a simple diode switch. The AC signal passes through C_1 and is stopped by D_1. This is because the mechanical switch (SW_1) is in the "off" position, supplying a negative, or reverse, DC bias to the diode. To forward bias the diode and allow the signal to pass, the switch is turned to the "on" position, which places a positive voltage that forward biases the diode into conduction, allowing the AC signal to ride on this DC bias. C_2 blocks the DC bias and passes the AC signal. R_1 is a current limiting resistor to ground for the two DC bias voltages, while the RFC blocks the RF from entering the power source. Using a diode instead of a simple mechanical switch likewise allows for remote switching of an RF signal without the added inductance and stray capacitance of a wire run to a distant panel switch.

Appendix F—Important Formulas

$V = IR$

$I = V/R$

$R = V/1$

$P_{TOTAL} = P_1 + P_2 + P_N$

$P = IE$

$P = I^2R$

$P = E^2/R$

Total impedance of a circuit: $Z_{TOTAL} = V_{APPLIED} / I_{TOTAL}$

Resonant frequency of an LC circuit: $fr = \dfrac{1}{2\pi\sqrt{LC}}$

$f = 1/t;\ t = 1/f;\ \lambda = C/f;\ f = C/\lambda$
(Where f = frequency in Hertz, t = time in seconds, λ = wavelength in meters, C = speed of light in meters.)

RMS (effective value) = .707 x V_{Peak}

V_{Peak} = 1.414 x RMS

$V_{Average}$ = .636 x V_{Peak}

To find instantaneous voltage anywhere on a sine wave, if degree of rotation is known:
$$V_{INSTANTANEOUS} = \text{SINE OF DEGREE OF ROTATION x } V_{Peak}$$

Time that one wave leads or lags a second wave: $t_L = \theta/360 \times 1/f$

kWh = kW x Hours

Voltage divider formula: $V_{R1} = R_1/R_T \times V_T$

Current divider formula: $I_{R1} = \dfrac{R_2}{R_1+R_2} x I_t$

For transformer turns ratio: N_P/N_S

Matching transformer turns ratio needed for impedance match: $\sqrt{\dfrac{Z_P}{Z_S}}$

Series inductances: $L_T = L_1 + L_2 + L_N$

Parallel inductances: $1/L_T = 1/L_1 + 1/L_2 + 1/L_N$

Series inductive reactance: $X_{L\ TOTAL} = X_{L1} + X_{L2} + X_{LN}$

Parallel inductive reactance: $1/X_{LTOTAL} = 1/X_{L1} + 1/X_{L2} + 1/X_{LN}$

Inductive reactance: $X_L = 2\pi fL$

Series capacitance: $1/C_T = 1/C_1 + 1/C_2 + 1/C_3 +$ Etc. (Equivalent to thickening, or spacing out, the dielectric.)

Parallel capacitance: $C_T = C_1 + C_2 + C_n$

Series capacitive reactance: $X_{CT} = X_{C1} + X_{C2} + X_{CN}$

Total parallel capacitive reactance: $1/X_{CT} = 1/X_{C1} + 1/X_{C2} + 1/X_{CN}$

Reactance of a capacitor at a desired frequency: $X_C = 1/2\pi fC$ (Applies only to sine waves.)

$Q = $ fr$/BW$ at-*3dB* or $X_L/$Re

Q in a series resonant circuit: $Q = V_{OUT}/V_{IN}$

Q in a parallel resonant circuit: $Q = Z_T/X_L$

$BW = $ fr$/Q$

Time it takes for current to rise to 63.3% of its steady state value in a resistive/reactive circuit: $t = L/R$ and $t = RC$.

Total current in a parallel resistive/reactive circuit: $I_T = \sqrt{I_R^2 + I_L^2}$

Reactive power: *SINθ(VI)* (Measured in VARs (*volt amp reactive*))

Apparent power: *EI* (Measured in VA (*volt amps*))

Real power (RMS) in watts $= I^2R$

Power factor: I^2R/IE or $POWER_{REAL}/POWER_{APPARENT}$ (PF will always be less than 1.)

PEP (AM) $= 4 \times P_{CARRIER}$

Transistor output power: $P_{OUT} = I_C \times V_C$

Common emitter amplifier gain calculations: $A_V = V_{OUT}/V_{IN}$; $\beta = I_C/I_B$; $A_P = P_{OUT}/P_{IN}$ or $\beta \times A_V$ (Where β = current gain, A_V = voltage gain, and A_P = power gain.)

FET amplifier gain: $G_m \times Z_{OUT}$

AC current gain: $\beta(h_{fe}) = \Delta I_C/\Delta I_B$ in a CE amplifier

DC current gain: $\beta_{DC}(h_{FE}) = I_C/I_B$ in a CE amplifier

Input resistance for CC amplifiers: $R_{IN} = h_{fe} \times R_L$ or $\beta \times R_L$

Amplifier efficiency: $RF_{POWER\ OUT}/DC_{POWER\ IN} \times 100$

Power supply regulation: % REGULATION $= V_{NO\ LOAD}\text{-}V_{FULL\ LOAD}/V_{FULL\ LOAD} \times 100$

$dB = 10LOG\frac{P_2}{P_1}$

$dBm = 10LOG\frac{P_2}{1mW}$

$dB = 20LOG\frac{\frac{V_2}{\sqrt{Z_2}}}{\frac{V_1}{\sqrt{Z_1}}}$ or, if Z_1 equals Z_2; $20LOG\frac{V_2}{V_1}$

Total AM power (with modulation): $P_T = \left(I_c\sqrt{\left(1+\frac{m^2}{2}\right)}\right)^2 \bullet Z_{OUT}$

(Where I_C = unmodulated carrier current, m = modulation index, and $Z_{OUT} = Z_{OUT}$ of final amplifier.)

FM BW: $2N \bullet f_{m\ max}$ (Where N = # of significant sidebands, and $f_{m\ max}$ = maximum modulating f[...] quency.)

FM modulation index (m): $f_{DEVIATION}/f_{AUDIO}$

FM deviation ratio: $f_{DEVIATION\ MAX}/f_{AUDIO\ MAX}$

ERP: $POWER_{TX\ OUT} \times (ANTILOG \frac{ANT\ GAIN\ IN+B,LESS\ LINE\ LOSSES}{10})$

RF power density: W/CM^2 (Watts per centimeter squared.)

Simple signal-to-noise ratio measurement: V_{SIGNAL}/V_{NOISE}

SINAD: *SIGNAL+NOISE+DISTORTION/NOISE+DISTORTION*

Measuring SWR with a wattmeter: $1 + (P_{REFLECTED}/P_{FORWARD}) \div 1 - (P_{REFLECTED}/P_{FORWARD})$

Damping factor: Z_L/Z_{OUT} (Where Z_L = Z of a speaker, and Z_{OUT} = Z_{OUT} of an amplifier.)

Common mode rejection ratio: $C_{MRR} = A_D/A_{CM}$ (Where A_D = difference gain, or 180° out-of-phase signals, and A_{CM} = common mode gain, or 0° in phase signals.)

Duty cycle: (%) = *PULSEWIDTH/PERIOD* x 100

Glossary

Absorption Trap — A type of wave trap that is inductively coupled to the output coil of an amplifier. It consists of a resonant tank circuit, tuned to the undesired signal, which absorbs the undesired frequency's energy.

Absorption Wave Meter — A tuned circuit with a tunable capacitor and a rectified meter. The coil of the tank circuit is loosely coupled to another tank or RF producing circuit. When the variable capacitor of the meter is varied a peak in the meter will indicate that the wave meter has coupled energy from the circuit under test, and thus that both are at the same resonant frequency. The resonant frequency is read from a dial or chart. Can be utilized to detect parasitics, to measure the frequency of an oscillator, or as a general purpose RF indicator.

Active Device — A component that generates or amplifies a current or voltage. Common examples would be transistors and vacuum tubes.

Automatic Level Control (ALC) — Controls the gain of a transmitter, controlling the drive to the final amplifier or some earlier stage. It can be used in AM, FM, and especially SSB transmitters. ALC will reduce excessively high amplitude output levels, but amplify medium and low level signals normally.

Alpha Cutoff Frequency (f_{ab}) — The frequency at which a common-base amplifier increases until the gain reaches 70.7% of its initial frequency of 1 kHz.

Amplifier Efficiency — The percent of RF output power compared to DC input power.

Angle-Of-Radiation — The angle of the transmitting antenna's radiation in relation to the earth's surface.

Automatic Noise Limiter (ANL) — Usually employing diodes, placed just after the detector to cut off any noise spikes that would reach the audio section of the receiver.

Apparent Power — The power that appears to be consumed in a reactive circuit, and is calculated by voltage x current (VI). Measured in VA, or Volt Amplifiers.

Application Specific Integrated Circuit (ASIC) — A custom designed and fabricated IC.

Attenuators — Either fixed or adjustable (step attenuators) apparatus for reducing signal amplitude, while maintaining the proper input and output impedances of any device attached to its ports.

Autopatch — A device for connecting a transceiver to the telephone lines by remote control.

Autotransformer — A single-winding transformer with only 3 leads. Cannot be used for isolation purposes.

Axial Leads — Leads on a component placed in-line but opposite one another. The ordinary arrangement for most resistors and signal diodes.

Backplane — An area where everything is connected; a common connection point (a "rats nest").

Balun — A wideband transformer that is capable of matching balanced line (twin lead, a dipole antenna) to unbalanced line (coaxial cable), as well as altering RF impedances.

Bandpass — The frequency limits between half-power points of a signal or filter. For example, a certain channel may have a bandpass of between 92.788 MHz and 92.800 MHz. Compare with bandwidth.

Bandspreading — The technique, either mechanically or electronically achieved, that allows the tuning of a receiver across a wide range, yet permits accuracy when tuning in a narrow frequency channel. A planetary drive is an example of mechanical bandspreading.

Bandwidth — The amount of frequency spectrum, in hertz, utilized by a filter or channel. For example, a certain channel has a bandpass between 92.788 MHz and 92.800 MHz: it thus has a bandwidth of 92.800 MHz minus 92.788 MHz, or 12,000 Hz.

Baseband Signal — The modulation signal, whether voice, music, or digital information.

Battery Ampere-Hours — The length of time a battery can supply a certain amount of current. For example, a rating of 120 ampere-hours means a battery will deliver 120 amps for 1 hour, or 10 amps for 12 hours.

Beam Width — The width of the major lobe of an antenna's output electromagnetic wave, in degrees, between the half-power points.

Binary-Coded Decimal (BCD) — A digital code that differs from binary in that a nibble represents only a single decimal. For example, decimal 12 to binary is represented by $12_{10} = 1100_2$; decimal 12 to BCD is represented by $12_{10} = (0001_2)(0010_2)$

Bit — Binary digit. A "0" or a "1".

Blocking Interference — A receiver that is overloaded by a very strong received signal, causing a reduction or total lack of output.

BNC Connector — A high frequency connector good up to 2 GHz. Both 50 and 75Ω versions are available. Used for test equipment and handheld vertical antennas in low power applications.

Broadband Power Amplifiers — Amplifiers employed in modern low and medium power transmitters that need not be tuned to transmit. Simply turning the tuning dial to change frequencies, or switching the band switch, is all that is needed. With this type of amplifier spurious signals and harmonics within this wide bandpass are not attenuated. However, the transmitters are far easier and faster to use.

Buffer Amplifier — Typically, a common-collector amplifier, which naturally has a high input impedance and low output impedance. The buffer is placed between an oscillator or amplifier and the load. Buffering prevents the load from drawing too much current from the previous circuit, which would load that circuit down. The buffer still allows the same signal voltage to be applied to the load.

Buffer Register — Temporarily stores a word during data processing.

Butterworth Characteristic — A description of a filter's type of response. The Butterworth has a very flat response with no peak within its bandpass.

Bypass Capacitor — A capacitor utilized to shunt AC around a component.

Byte — 8 bits (1 character)

Capacitance and Inductor Multipliers — Circuits that can create artificially high values of capacitance and inductances, at low frequencies, by the use of op-amps and low value capacitors.

Carrier Shift — In AM transmitters, the unequal increase or decrease in amplitude of either the negative or positive alternation of the modulated wave, which should always remain symmetrical. Positive carrier shift, which results in an increase of average AM output power, can be caused by parasitics, improper tuning of the finals, bad neutralization, or excess audio drive. Negative carrier shift, which results in a decrease in the average AM output power, can be caused by any distortion in the audio modulation, overmodulation, bad tuning of the resonant tanks, or bad regulation. Carrier shift may be monitored by a circuit built into many AM commercial transmitters.

Carrier — The frequency used to carry voice or data modulation from the transmitter to the receiver.

Ceramic Filter — A narrow bandwidth bandpass filter. Smaller and cheaper, but with a lower Q, and thus a wider bandwidth, than a crystal filter. Can be in a two terminal or three terminal package, with the third terminal placed at ground potential. The input frequency is inserted at the center terminal, with the output taken from the outer terminal of the three terminal device. In a receiver the ceramic (or crystal) filter will generally be located between the mixer and the first IF amplifier, and sets the selectivity of the IF.

Channel — A segment of the frequency spectrum that includes a carrier frequency with its attendant sidebands; a bandpass taken up by one modulated signal.

Characteristic Impedance — Also called surge impedance. The AC resistance of a transmission line as seen by the generator (such as a transmitter) and the equivalent value of the termination needed to make that line flat (no reflected waves). This value does not change with the frequency of the signal or length of line, and is a function of the physical size, spacing, and properties of the transmission line.

Chebyshef Characteristic — A description of a filter's type of response. The Chebyshef has a certain amount of ripple within its passband but has a sharp cutoff response.

Choke Action — As the frequency increases in an inductor, inductive reactance will also increase. A device referred to as a choke uses this action to provide extremely high impedance to AC or RF, while letting DC pass un-attenuated.

Circuit Sharing — Modern circuit design that gives joint use of certain circuits in a transceiver, such as the oscillators, filters, power supplies, frequency synthesizers, and amplifiers, to both the receiver and transmitter sections.

CMOS Logic — A logic arrangement employing Complementary Metal-Oxide Semiconductors. Uses both P and N channel MOSFETs with very low current consumption. However, they are susceptible to static discharge damage and are slower than TTL logic. Newer designs exploit Zener static protection. Power supply can be anywhere between 3 to 15 volts, which can also be unregulated.

Communication-Service Monitor — A consolidated test unit that may be used to service FM, AM, and SSB in the field. Contains a signal generator, frequency counter, modulated FM and/or AM generator, audio generator, and FM deviation meter.

Compandoring — In SSB, compressing speech amplitudes at the transmitter and expanding them at the receiver. Some SSB compandored transmitters send a 3 kHz pilot tone to permit rapid tuning of a mobile receiver.

Complex Wave — A wave made up of a fundamental (sinusoidal) wave and its harmonics. Square and triangle waves are common examples. These complex waves will be distorted by reactive components, due to the different capacitive or inductive reactances and phase angles affecting each of the harmonics forming the wave.

Compression — Loud sounds are made quieter, quiet sounds are made louder in an amplifier. Used to lower bandwidth requirements and remove any chance of overmodulation in a transmitter.

Connector, RF — A device used to obtain a permanent or temporary connection for transferring RF energy, at the proper impedance, from one circuit or cable to the next. See *Specific Connector*.

Converter (DC-to-DC Converter) — A circuit utilized to transform one DC voltage value into another DC voltage value. This is accomplished by "chopping" the DC input voltage so that it can be converted into another value by a built-in transformer. An output filter converts this chopped wave back into direct current.

Copper Losses — Losses in lines or coils caused by I^2R losses, skin effects, and crystallization (wire damage due to high temperature or bending by wind, causing small, high resistance cracks).

Coupling Capacitor — Couples AC to another stage, while blocking DC.

Cross-Modulation — Interference created when two powerful signals are adjacent in frequency. The desired signal is modulated by the undesired signal in the non-linearities of an amplifier in the receiver's front end, causing amplitude modulated interference, which can be heard in addition to the desired signal. The amplifier may only be partially nonlinear, or may be driven into nonlinear operation by the presence of a strong local AM broadcast station. May be reduced by traps and filters at the receiver's antenna input in order to filter out the unwanted frequency.

Crosstalk — Undesired electromagnetic or electrostatic transfer of signals between systems, channels, or sections of a system.

Crystal Oven — Employed in extremely accurate oscillators to stabilize the temperature of a crystal, which eliminates temperature induced crystal frequency drift. Consists of a small oven chamber maintained at a slightly elevated temperature over that of ambient by a heating element and internal thermostat.

Cutoff Frequency — The points on a signal's selectivity curve that are 0.707, or 3 dB (the half power points), below the signal's peak amplitude. Regularly used to measure the bandwidth of a tuned circuit.

Damping Factor — A numerical value of an amplifier and speaker impedance combination. The lower the output impedance of a power amplifier and the higher the speaker input impedance, the cleaner and sharper the sound output of an audio amplifier.

Darlington Pair — Twin transistors directly coupled, with a very high input impedance and a very high gain.

Decoding — An electronic digital function of converting from binary to decimal.

Desensitization (Desensing) — A close unwanted frequency can create an overloading in a receiver's first RF amplifier or mixer, causing a loss of sensitivity to the desired signal.

Degenerative Feedback — 180° out-of-phase feedback from an amplifier's output back into its input. This action lowers both its gain and its distortion levels. Can be designed into a circuit by utilizing feedback-producing components. Or it can be undesirable feedback if it occurs due to the interelectrode capacitances of the active amplifying device. Also called negative feedback.

DIAC — A thyrister employed as triggers for TRIACs to equalize the ON voltage for each AC alternation.

Die Form — An IC chip used without a package to conserve space.

Direct Conversion — In a receiver, converting the modulated RF signal directly into an audio frequency, which is then detected and fed to audio amplifiers; no IF is utilized.

Directional Coupler — A device used to sample the power traveling down a transmission line in one direction, while attenuating the signal in the other direction, thus not allowing any reflected waves back down the line, significantly lowering SWR.

DMOS — Method of producing a high power, high current, high gain, high frequency RF power MOSFET.

Disc Operating System (DOS) — A system program that informs a computer how to operate the disc drive, set up files, erase files, find files, etc.

Double-Tuned Transformer — A transformer that contains two slugs; one tunes the primary, the other tunes the secondary.

Downlink — The frequency utilized by a satellite to send signals to earth.

Driver Stage — The stage that drives the power amplifier. A pre-driver drives the driver.

Digital Signal Processing (DSP) — A incoming signal is converted to digital pulses, manipulated, then converted back to an analog waveform. Used in digital state-of-the art adjustable filters. DSP improves S/N ratios and fidelity.

Dummy Load — A load for transmitters that simulates an antenna, but does not cause radiation. Used for testing and adjustment.

Duplexer — A device that allows one antenna to be operated for both transmitting and receiving.

Duty Cycle — Ratio of a square wave to its ON time and OFF time.

Dynamic Instability — Undesired changes in a transmitter's frequency.

Dynamic Range — The measurement, in decibels, between the faintest and loudest sounds reproduced without significant noise or distortion in a system.

Eddy Currents — Induced currents in the core of a transformer with resulting I^2R (power) losses given up as heat. The higher the frequency the greater these losses become. Eddy currents can be reduced by laminating the core with thin sheets instead of one solid core, or with the use of ferrites or air cores.

Electronics Industries Association (EIA) — Sets equipment and other standards in the electronics field.

Electro-Magnetic Interference (EMI) — Interference to any electronic device caused by a producer of electromagnetic energy, such as transmitters, computers, electric motors, relays, etc.

Encoding — An electronic digital function of converting from decimal to binary.

Effective Radiated Power (ERP) — ERP is the actual power delivered to the antenna times the power gain of the major lobe of the antenna.

Exciter — The carrier oscillator for a transmitter, or the driven element of a directional antenna.

Fading — The intermittent decline of a received signal in a receiver. Can be caused by out-of-phase signals reaching the receiver's antenna due to multi-path reception, or changing conditions in the ionosphere.

Fan-Out — Maximum number of digital gates one gate can drive.

Faraday Shield — Also referred to as an electrostatic shield. A shield placed between two coils, such as the primary and secondary of an air-core transformer in an RF tank circuit, that drastically reduces the natural capacitance that occurs between two conductors separated by a dielectric (in this case, air). Harmonic and spurious signal frequency transfer between coils are almost eliminated, and are passed to ground. The shielding allows the resonant frequency to be coupled to the secondary through transformer action, with no adverse effects on the collapsing and expanding magnetic fields that cut the conductors of the secondary.

FCC Type Acceptance — A type acceptance number given by the FCC to a manufacturer for a device that meets certain frequency, bandwidth, and stability requirements. If a non-original or non-approved part is used, type acceptance could be nullified, making the device illegal.

F-Connector — Common RF connector used in TV receiver connections and very low power applications.

Feedback Amplifier — An audio amplifier that exploits a negative feedback loop to supply degenerative feedback in order to minimize distortion. However, this will lower overall stage gain. Distortion in the feedback loop may be quite high.

Feedthrough Bypass Capacitor — A capacitor that passes DC but sends AC to ground. Commonly used at the input to a shield or metal cabinet containing a transmitter, receiver, or power supply.

Field Strength, Antenna — A measurement, generally in μV/meter, of the electromagnetic energy radiating from an antenna.

Flip-Flops — Digital two-state memory elements.

Flywheel Effect — The ability of a tuned circuit to complete a sinusoidal wave if only a pulse is received. The capacitor and inductor of a tank circuit exchange energy back and forth, causing oscillation at the tank's resonant frequency. Just as long as a pulse is received at set times to re-enforce and replace the power lost in the unavoidable resistances present, the circuit will continue to oscillate, forming nearly pure sinusoids.

Four Layer Diode — Semiconductor device found in overvoltage indicators, relaxation oscillators, and time delay circuits. Four layer diodes are turned on by a trigger pulse, and can be utilized for switching on SCRs.

Frequency Shift — An unwanted shift in oscillator frequency. Usually caused by a low C/L ratio, with the resulting low Q, in the tuned circuit of the oscillator; or by loading down of the oscillator; or by poor power supply regulation.

Frequency Synthesis (FS) — A technique for generating extremely accurate multiple frequencies from one (but generally more) crystals.

Frequency Tolerance — The amount, in a percentage, that the carrier wave is allowed to deviate from its center frequency.

Frequency Translation — Changing one frequency to a higher or lower frequency, while still retaining the original modulation information.

Full Duplex — Both ends of the communication channel can communicate with each other during the same time frame.

Fundamental Frequency — The frequency at which a pure sine wave's energy is located. Except in frequency multipliers (harmonic generators), it is the frequency that the tuned circuits of RF devices are tuned to.

Fundamental Overload — Interference to an electronic device, such as a radio or television receiver, caused by the fundamental frequency of a transmitter reaching the receiver's amplifiers and producing undesired results. The receiver is incapable, through careless or faulty design, to reject out-of-channel signals.

GASFET — A gallium arsenide FET used at UHF and above.

Grid Dip Meter — Basically an oscillator with plug in coils for gross adjustment and a variable capacitor for fine tuning. When the frequency of the grid dip meter is adjusted to the resonant frequency of a circuit's tank, the energy of the meter is coupled to the tank, lowering the amplitude of the oscillator's oscillations, thus causing a dip in the meter. Can be utilized to measure the approximate resonant frequency of any tuned circuit.

Ground Fault — Loss of ground, caused ordinarily by a severed ground strap or wire, or a bad ground connection. Also refers to any alternate return path to ground for current, such as through the human body.

Half Duplex — Each end of the channel can communicate with the other, but not at the same time.

Handy-Talkie — A walkie-talkie. Customarily referred to simply as an HT.

Hardware — The physical computer equipment, such as the CPU or the hard drive.

Harmonics — The frequencies generated by a distorted sinewave at multiples of its fundamental. The harmonic's frequencies can be at two, three, or four or more times that of the fundamental. If it could be attained in practice, a pure sine wave would have no harmonics, but would consist merely of its unadulterated fundamental frequency.

Harmonic Distortion Analyzer — A test instrument used to measure total harmonic distortion (THD). To use: 1. Inject an audio frequency into the receiver input at the proper amplitude. 2. Set selector to broadband, then feed total signal into input, while setting meter to 100%. 3. Switch in the fundamental rejection filter and take reading (meter is calibrated in % THD).

Harmonic Generator Frequency Interference — Any harmonic of the fundamental frequency generated from a transmitter, frequently causing interference, such as from the master oscillator or frequency doublers or triplers. Any of these can leak from the transmitter's shields or cabinet, creating interference.

Harmonic Suppression — The degree to which harmonics at the output of a transmitter are attenuated.

Helical Resonators — Basically a helically wound resonant transmission line surrounded by a conductive shield. Can be operated between 50 MHz and 500 MHz. They are employed as highly selective filters at the input to high frequency receivers. The resonator can be tuned by an air dielectric variable-capacitor or by tuning screws.

Helix — A single layer coil.

HEXFET — A low drive current, low resistance, high gain, temperature stable metal-oxide semiconductor field-effect transistor, used typically in switching power supplies and audio amplifiers.

Hook-Up Wire — Wire utilized to connect circuits or equipment.

Hybrid Combiner — A device that allows two signal generators to inject their signals into a receiver without adversely affecting each other.

Hysteresis Loss — Losses in a transformer's core created by friction of the shifting magnetic domains in the ferromagnetic core, caused by the alternating current signal, producing heat and a resulting power loss. Air-core coils have almost no hysteresis loss.

Incidental AM (IAM) — Unwanted AM of an FM signal.

Incidental FM (IFM) — Unwanted FM of an AM signal.

Inverter — A circuit used to convert DC into AC.

IRED — An infrared LED.

Isolator — A device constructed of ferrites, generally used at microwave frequencies, that allows a signal to pass in one direction, but not in the reverse direction. It is a one-way device that can be employed to prevent reflected waves or externally generated frequencies from entering the finals of a transmitter.

International Telecommunications Union (ITU) — A coalition of 154 member countries that negotiate worldwide RF spectrum allocations.

Junction Tetrode Transistor — A four terminal high frequency transistor with two base leads: one grounded, the other at 0.4V. This reduces base resistance and increases its upper frequency limit.

Knife-Edge Diffraction — When an electromagnetic wave reaches the top of a barrier (such as a mountain), it has the tendency to bend slightly, causing reception in an area were it would not normally be possible.

Liquid Crystal Display (LCD) — When electricity is applied to liquid crystals the molecules of the crystals become opaque.

Lead Acid Cells — A common high current backup battery that is rechargeable, with a voltage per cell of 2.1V and a specific gravity of 1.27 at full charge.

Lecher Lines — Used to measure wavelength, and thus frequency. They consist of a transmission line with adjustable sliding contacts. When these contacts reach the standing waves, the distance is measured between the voltage peaks or valleys, thus obtaining its wavelength.

Linear Mixing — Linear mixing creates one frequency that rides on another, usually creating hum. Unlike nonlinear mixing, no extra frequencies are produced.

Litz Wire — Specially woven insulated multi-strand wire capable of high frequency operation with low losses.

Loading Coils — Electrically increase the length of an antenna to make it appear longer than it actually is, decreasing its resonant frequency. However, they will also narrow the effective bandwidth of an antenna. (Adding a capacitor in series with an antenna will increase its resonant frequency, or electrically shorten the antenna).

Local Area Network (LAN) — A connection, by wire or radio, of computers over a small area, commonly confined to a single building or group of buildings, to share data or peripheral equipment.

Locking — Locking two oscillators together refers to electronically synchronizing the frequency of one oscillator to the frequency of another.

Logarithmic Taper — The resistance change of a potentiometer that varies logarithmically to simulate the response of the human ear, in contrast to linear tapers. Operated as audio potentiometers.

Microwave Integrated Circuits (MIC) — ICs with components integral to a small stripline or microstrip sections.

Microcontroller — A circuit, placed on a single chip, that contains a CPU, RAM, an EPROM, I/O ports, timers, and an interrupt controller.

Microphone— A transducer that converts sound waves into current. There are many types: ceramic and crystal mikes have a flat frequency response, require no power supply, have a high impedance output, but have low output levels, with some hum pickup; dynamic mikes have a moving coil with low impedance output, require no power supply, have good frequency response and high output levels, with low hum pickup and large dynamic range; ribbon mikes have a wide frequency response, high output level, with no power supply necessary, have low impedance output, a large dynamic range, and low hum pickup; condenser mikes use a "voice" variable capacitor, have low output levels, and many need external power supplies.

Microphonics — Physical vibration causing modulation of a receiver or transmitter. Created by loose components, especially in the oscillator regions. More common in tube equipment due to their sensitive electrodes.

Microprocessor — A CPU on a single chip. Frequently called the MPU (microprocessor unit).

Modulation — Combining a lower frequency, the modulation, with a higher frequency, the carrier. The modulation frequency may be voice or data.

Modulus — The modulus of a digital counter is the number of counting states before it begins to repeat itself.

Monolithic-Crystal Filter — A small, low cost, multielectrode crystal filter used for frequencies from 5 MHz to 300 MHz.

Motorboating — Unwanted self-oscillations in an audio amplifier causing a motorboat type "putt-putt" noise. May occur intermittently.

Multiple-Path Reception — A receiver receiving both ground wave and sky wave propagation, or by one-hop and two-hop paths, or by sky wave and surface wave. The two signals are alternately inphase and out-of-phase with each other, which creates fading.

Multiplexer — A device with many inputs but only one output. By using control signals, any input can be steered to the single output.

Multivibrators — Circuits that utilize two transistors that alternate between saturation and cutoff. MV's are pulse generators of square or rectangular waves and are composed of four basic types. **A**. STABLE MV: Sends pulses only when triggered; **B**. ASTABLE MV (free running): Sends pulses constantly without a need for triggering; **C**. BISTABLE MV (flip-flop): One input causes one 180° phase shifted pulse out; **D**. MONOSTABLE MV (one shot): Causes one output pulse of any preset length for each input pulse.

N-Connector — A 50Ω RF low reflection connector good up to a maximum of 18 GHz. A 75Ω version is available, but the two are not compatible. Used for test equipment hookups and for coaxial and chassis connections.

Network — A circuit grouping of electronic components. An example is the RF RC "π" network that can transform one impedance to another by modifying the reactive component and changing the circuit resistance to a desired value. Also refers to a group of interconnected receiving, transmitting, or transceiving stations.

Noise Figure (NF) — The degree that a receiver approaches the quietness of a theoretical receiver. In the real world, 4 dB is superior.

Notch Filter — A sharp band-stop filter.

Non-Return to Zero (NRZ) — A method of decreasing bandwidth requirements for data transmission by lowering the number of times the signal changes states, by forming any serial string of incoming "1"s as an unchanging HIGH state — until a "0" occurs, in which case the level drops to a LOW until the next HIGH occurs. (See *RZ*).

Octave — A single octave is twice a given frequency or half a given frequency.

Original Equipment Manufacturer (OEM) — Generally used when referring to original replacement parts for the device under repair.

Opto-Isolator — A device that utilizes an LED and a photodiode or photo-transistor (an optical coupler) to completely isolate one circuit from another, with no noise feedback or loading on the first circuit. The signal is fed into the input, converted to light by the LED, sent through a small gap, and focused into a light sensitive device that converts the light beam back into a voltage.

Optoelectronic Devices — Devices that either generate or control current by the application of light energy, or devices that generate light by the application of current. Examples are photoresistors (photocells), ordinarily made of CdS (Cadmium Sulfide), that vary resistance (between 300Ω to $200\ \text{M}\Omega$) when light is applied or removed. Have slow response times, a memory effect and a nonlinear response to changes, but are very sensitive to light variations. Photodiodes are used as photoresistors. Phototransistors are operated as high sensitivity photodiodes, but have higher current carrying capabilities, but at slower response times. Photovoltaic, or solar cells, when exposed to light, generate a voltage. When a photon hits the cell an electron can be dislodged from the atom's valence, thus causing electron flow. They have low sensitivity, low open circuit voltage, and low closed circuit current. LEDs generate light energy when a voltage is applied to their leads. Typical values of current would be 50 mA with a forward voltage of 1.6V.

Pad Attenuator — Reduces the output amplitude of an input signal, while matching the input and output impedances of the circuits over their complete frequency range.

Padder Capacitor — A trimmer capacitor for small levels of low end frequency adjustment in a superheterodyne's local oscillator.

Padder Trimmer — A trimmer capacitor for small levels of high end frequency adjustment in a superheterodyne's local oscillator.

Parameter — The electronic characteristics of a device.

Parametric Amplifier — Also known as para-amps. High frequency amplifiers (primarily 30 GHz and above) with the traveling wave tube or varactor operated as the para-amp's main element. Very low noise, which can be further reduced by cryogenic cooling.

Parts-Per-Million — A ratio measurement of the purity or accuracy of a device. For example, a certain frequency counter is specified as being accurate to within $\pm$ 0.1 PPM. It reads a transmitter's output as 156.52 MHz. Thus, the reading of the counter will differ by a maximum of $\pm$15.652 Hz from the transmitter's actual frequency, or ($0.1/10^6$ 156.52 MHz = 15.652 Hz).

PEP — The peak power of an AM or SSB envelope with 100% modulation applied to the carrier. $PEP = V_{RMS} \bullet I_{RMS} = V^2_{RMS}/R$

Periodic Waveform — A wave that does not change from one cycle to the next. It can be a sine, square, sawtooth, etc., waveform. The sine wave is the only periodic waveform that cannot be distorted by RC, LC, or RL networks.

Peak Inverse Voltage (PIV) — The maximum reverse voltage a diode can sustain before breakdown occurs.

Phase Modulation (PM) — A technique of producing indirect FM.

Point-to-Point Wiring — An old method of circuit wiring employing wire to connect components, instead of utilizing the more modern printed circuit board traces to conduct current. Still used in some high current applications and in wire-wrap circuits.

Pole — In reference to filters, a single pole is an increase in signal attenuation of 6 dB as the frequency doubles or is cut in half (one octave) at the filter's resonant frequency on the frequency response curve. Pole also refers to the number of crystals in a filter: a two-pole filter has two crystals. The more poles, the sharper the filter's skirt and the faster the drop on the response curve.

Port — The point or terminal in a circuit or device that a signal may be injected or extracted.

Potentiometer — An adjustable device that controls voltage. Also an old term occasionally used to describe voltmeters.

Premixer — A technique of obtaining a stable higher variable frequency by the nonlinear frequency mixing of a stable crystal oscillator with that of a variable oscillator.

Power Supply Unit (PSU) — A term for a power supply circuit.

Q-Point — The area, in an amplifier, where the DC operating bias is set when no signal is present. Depending on the location of the Q-point, an amplifier can operate class A, AB, B, or C. Also called the quiescent, or resting, point.

Quadrature Phase — Two sine waves that are 90° out of phase with each other.

Quasi-Complementary Symmetry — A complementary symmetry amplifier that employs two NPN Darlington pairs instead of one NPN and one PNP transistor. The two NPN Darlington transistors are lower in cost and easier to match than the transistors in a true complementary symmetry amplifier.

Radial Leads — Leads on a component in parallel with one another. The common lead arrangement for capacitors.

Radials — Also referred to as a ground plane or counterpoise. Used when insufficient conditions exist for grounding a quarter wavelength vertical antenna. Consists of four or more quarter-wavelength wires laid on or in the ground, to simulate a proper earth ground.

Radiation Resistance — The AC resistance of an antenna at its feedpoint.

Rails — The long thin plastic packaging for storing ICs, or the maximum negative and positive voltages in a comparator.

Random Access Memory (RAM) — Memory that can be read and written from.

Reactive Power — An indication of the apparent power used in the reactive component of a circuit. Calculated by $SIN\theta$ (VI). Measured in VARs, or Volt Amps Reactive.

Reactor — A very high Q coil utilized principally for its high reactance.

Real Power — The only true power consumed in a reactive circuit is the power dissipated in the resistance, while the reactive components do not consume power, they store it. Real power is calculated by I^2R.

Regenerative Feedback — Amplifier feedback that is inphase from its output back into its input. This feedback can create increased amplification or oscillations, either desirable or undesirable. Also called positive feedback.

Register — A data storage circuit.

Reflectometer — Also called an SWR bridge. Employed to measure the standing-wave ratio of a transmission line.

Reluctance — Magnetic resistance.

Repeater — A mechanism that retransmits a signal that it receives from a transmitter, generally concurrently. Used to extend the range of a low powered transmitter or to lengthen the line-of-site of a low lying antenna.

Resistor Array — Multiple resistors presented in a single IC DIP (Dual Inline Package) or SIP (Single Inline Package).

Resonator — A device that, at its natural resonant frequency, vibrates, or oscillates, at this frequency, and will resonate at no other injected frequency. Adopted instead of standard LC circuits at higher frequencies for filtering.

Radio Frequency Interference (RFI) — Interference to any electronic device caused by a producer of RF, such as a transmitter, a computer, or the LO in a receiver.

Relative Field Strength Meter (RFS) — A field strength meter that employs relative, and not absolute, readings.

Rheostat — An adjustable device that controls current.

Ringing — Damped sine wave oscillations occurring when a tank circuit is hit by a pulse.

Read Only Memory (ROM) — Memory that can be read from, but not written to.

Return-to-Zero (RZ) — In the digital RZ format for data transmission, each "1" bit in an incoming serial data stream changes the RZ's signal state to a HIGH pulse, while any incoming "0" bits are held continuously LOW until an incoming "1" again occurs. (See *NRZ.*)

Sampling Scopes — Oscilloscopes that sample waveforms at repetitive time slices, then reassemble the waveform and display it on the CRT. Since only discrete samples are made, this can extend an oscilloscope's frequency response by 100 or more (for repetitive signals only).

Schmitt Trigger — A circuit that can receive an incoming ragged pulse and convert it into a clean square wave.

Selective Calling — Tone-coded signals are sent to receivers from a transmitter, opening the squelch of only a specific receiver, thus opening a communications channel.

Selective Fading — Fading that strikes at certain frequencies, commonly caused by multi-path transmission or single, double-hop transmissions reaching the receiver's antenna out-of-phase at times, inphase at others.

Selectivity Factor — A filter measurement that describes the steepness of its skirt, or *(BW)-3db/(BW)-60dB*.

Selectivity — The ability of a receiver to choose one signal out of many while rejecting all others. The receiver's bandwidth must be narrow enough, and with sharp enough skirts, but be wide enough to not attenuate the sidebands of the incoming signals. Generally the filters in the IF control this. Placing filters in series (cascading) narrows bandwidth, thus increasing selectivity.

Self-Resonance — A point is reached when a specific frequency causes a component to become resonant due to its distributed and stray capacitive and inductive reactances.

Sensitivity — The ability of a receiver to pick up low amplitude signals. A function of the gain, as well as the internal noise, of the receiver. Commonly expressed as the minimum signal that is 10 dB above the noise. Sensitivity is greatly influenced by the gain of the IF amplifiers and the noise generated in the RF and mixer sections. The best communication receivers can receive signals as low as 0.2 µV, while low cost AM consumer receivers may not be able to receive anything less than 75 µV.

Shot Noise — Random noise created by electron movement in a semiconductor or vacuum tube, resulting in a "hissing" sound.

Shrink-Fit Tubing — Plastic tubing placed over a wire or connection that, when heated, shrinks to tightly conform to the conductor. Performs the function of insulation.

Sideband Cutting — The higher audio frequencies become distorted due to the bandwidth of the IF filters being too narrow, and attenuating the higher sideband frequencies in a receiver.

Signal Diode — A small current diode used for AM detection.

Signal Monitor — Low cost spectrum analyzer which monitors a narrow band of signals, usually those in a receiver's IF.

Signal-to-Noise Ratio (S/N) — The ratio, in dBs, between the signal voltage or power, to the noise voltage or power.

Signal Tracer — A piece of test equipment, operated in signal tracing, that detects the audio modulation on the injected RF, and amplifies this audio so it can be heard by the operator.

Simplex Operation — One side of a communications channel can only communicate with the other, but never visa-versa. In voice communications, it refers to the alternating communications between two transceivers on the same channel, but not at the same time, without the use of a repeater.

Single-Ended — A one-transistor amplifier.

Skin Effect — A consequence of high frequencies, creating an increase in a conductor's effective AC resistance. The RF begins to travel closer to the surface as the frequency is raised, until finally conduction takes place only on the conductor's surface. This causes a higher value of AC resistance than that measurable by a normal ohmmeter. The skin effect also produces a decrease in the Q of a coil.

Skip Zone — Also called the zone of silence. An area were no reception is possible. It occurs at the end of ground wave propagation, but before the sky wave's return to earth.

SMA Connector — A common microwave connector good up to 24 GHz. Generally used for semirigid cables and components that will be semi-permanently connected.

Surface Mount Devices (SMD) — ICs and discrete components operated in surface mount technology (SMT) applications.

Surface Mount Technology (SMT) — A method of using special ICs and discrete components that are not soldered in through-holes in the PCB board, but are soldered to the board's surface.

Software — A computer's detailed digital programs for performing a desired function.

Small Outline Integrated Circuit Package (SOIC) — An SMD IC used in surface mount technology applications.

Spaghetti — Hollow insulating sleeves for un-insulated wire.

Speech Compression — A form of AGC in the audio stages of an AM and SSB transmitter. Speech compression maintains the average power output at an increased level by amplifying low amplitude audio more than higher amplitude audio.

Speech Processor — A circuit that controls the amplitude and/or frequencies of a modulating speech input to a transmitter's modulator. Suppresses overmodulation caused by excessive speech volume, as well as excessive bandwidth created by high frequency audio.

Splatter — A collection of spurious sidebands caused by overmodulation.

Spurious Emissions (spurious radiation or "spurs") — Interfering emissions that are outside of the design bandpass of a transmitter.

Step Attenuator — A device that allows adjustable attenuation from its input to its output, in discrete dB calibrated steps, while matching impedances. (Input and output impedances are customarily 50Ω). Utilized to lower the output power of transmitters and frequency generators to safe levels for insertion into a spectrum analyzer, frequency counter, etc.

Sweep Generator — A frequency generator that can sweep through a band of frequencies, with variable sweep widths and output amplitudes. Employed to test the frequency response of amplifiers and filters in conjunction with an oscilloscope.

Telemetry — The transmission of measurement signals remotely, such as pressure, positions, frequency, temperature, speed, etc.

Temperature Compensated Zeners — Zeners that do not have large changes in Zener voltage (V_Z) with large changes in temperature.

Termination — A load used to terminate an output circuit, customarily in its characteristic impedance, to avert damage or unexpected operation. Prevents reflected waves.

Thermal Noise — Noise created in an amplifier by the random movement of electrons (white noise).

Thermal Runaway — A condition created in a semiconductor by increasing heat causing increased current, which produces further heat until, unchecked, the common result is destruction of the semiconductor device.

Thermistor — A temperature-sensitive resistor that changes its resistance in a nonlinear fashion as the ambient temperature changes. Thermistors may have either a positive or a negative temperature coefficient.

Thermocouple — A device constructed of two dissimilar metals that produces electricity by the application of heat to these metals, causing a difference in potential.

Thyristors — Solid-state electronically controlled switches. They may also be used to control the amount of power reaching a load. SCRs, TRIACs, and DIACs are common examples of thyristors.

TNC Connector — An RF connector good up to 12 GHz. Similar to the BNC, except threaded connections are employed instead of a bayonet.

Tracking — The ability, in a superheterodyne receiver, for the local oscillator to maintain the intermediate frequency (IF); or for the RF and IF tank circuits to mutually remain on-frequency when utilizing ganged capacitors.

Transient Intermodulation — Distortion caused by an amplifier's inability to respond to an input signal's fast changing amplitude variations.

Transient Voltage Suppressors (TVS) — Devices, typically Zener diodes, maximized for use as transient voltage suppressors in sensitive circuits.

Transit Time Problem — In a tube, as an electron is "boiled" off of the cathode, it begins to travel toward the control grid, heading toward the plate. Before this electron can pass the control grid, at higher frequencies, the control grid has already begun to turn negative, and now repels the electron back toward the cathode. Lowers gain and causes transit time noise. Transistors are also susceptible. (See *Frequency Limitations*.)

Translator — A type of repeater operated to retransmit TV signals into inaccessible areas.

Transponders — Satellite-based transponders are broadband repeaters that can retransmit many channels at once, while non-satellite transponders automatically broadcast an ID signal in response to another transmitter's interrogation signal.

Trap Antennas — A type of multiband antenna capable of operating on different bands due to the presence of several resonant tuned tanks along its length. Trap antennas can radiate harmonics.

Trimmer — An adjustable fine-tuning resistor, capacitor, or inductor of small values and size.

Trunked RF Systems — Consist of a base station that can communicate with a large number of portable units. Generally in the 806 to 866 MHz range, with 5 to 20 channels available and serving up to 2500 radio units.

Thin Small Outline Package (TSOP) — IC employed in surface mount technology applications.

Transistor-Transistor Logic (TTL (T²L) Logic) — A digital logic circuit consisting of transistors. Its power supply must be a well regulated +5 volts.

Television Interference (TVI) — Caused by a producer of RF, such as a transmitter, computer, or the LO in a receiver.

Twisted Pair — A transmission line consisting of two insulated wires twisted together. Employed for short runs in low frequency applications to reduce low frequency magnetic hum pickup and EMI producing radiated fields. Characteristic impedance is typically about 100Ω. There are two types; shielded and unshielded.

Temperature-Compensated Crystal Oscillators (TCXO) — Oscillators capable of supplying very stable reference frequencies. Commonly utilized in PLL circuits.

UHF Connector — Also referred to as a PL-259 connector. Used with medium to low RF power levels at 200 MHz and below, with 50Ω impedances. A very common and mechanically strong connector.

Unijunction Transistor (UJT) — Three-terminal, single-junction electronic device operated as a switch in UJT relaxation oscillators, which produce certain types of non-sinusoidal waveforms, as well as a trigger in SCR/TRIAC circuits; and as a switch. A PUT is a programmable UJT.

Uplink — The frequency employed to send a signal up to a satellite.

Variable Crystal Oscillator (VXO) — A crystal oscillator that uses a variable capacitor or inductor in series or parallel with the crystal to obtain very limited frequency adjustment capabilities.

Varistor — Consists of the MOV (Metal Oxide Varistor) and the ZOR (Zinc Oxide Resistor), and are used to protect circuits against transients: When a spike of voltage reaches a certain level, the varistor's resistance goes down, shunting the transient to ground.

Vias — Through-hole vias are plated hollow connections placed between tracks on opposite sides of a double-sided (tracks on both sides) PCB, or completely through a multi-layer PCB from top to bottom. Blind vias connect one exposed layer of a multi-layer board to a hidden lower layer.

Virtual Ground — A point in a circuit that appears to be at ground, but is not actually attached to ground. For example, since the input to an op-amp has such an extremely high input impedance, very little current (virtually zero) flows into an op-amp, making the input a virtual ground because zero current equals zero volts, which equals ground potential.

VMOS — An older method of producing a high power, high current, high gain, high frequency RF power MOSFET. Replaced by DMOS.

Voltage Dependent Resistor (VDR) — A device that decreases resistance continuously as the voltage rises. Can be used in current regulation circuits.

Wave Analyzer — Essentially a frequency selective voltmeter.

Wavetrap — Also called a bandstop filter. Designed to eliminate a certain frequency from entering a receiver or exiting a transmitter.

Word — A group of bits filling a single storage location in a computer. May consists of 8 bits, 16 bits, 32 bits, etc.

Zero Beat — When two signals are mixed together in a nonlinear fashion, both the sum and difference of the two signals are created. If the difference frequency is sent to a speaker, then as the two original signals are brought closer together in frequency, an audio tone can be heard. Since the difference between these two signals decreases as they become closer in frequency, a point is reached were there will be no difference in frequency between the two signals. This is a null point, where no audio tone can be heard at all. This null point is referred to as "zero beat".

Zone of Silence — See *Skip Zone*.

BIBLIOGRAPHY
Recommended Reading

The following are some of the finest books in the electronics field for technicians. A few may be out of print, while others may be published in newer editions than those shown below.

The ARRL Handbook for Radio Amateurs
Schetgen, Robert (Editor)
ARRL, 1995
ISBN 0-87259-172-7
Published yearly. Contains beginning, intermediate and advanced RF troubleshooting, construction, theory, and design information.

Basic Electronics
Grob, Bernard
McGraw Hill, 1984
ISBN 0-07-024928-8
A very informative book for those beginning in electronics.

Basic Electronics
The U.S. Navy
Dover Publications, 1973
ISBN 0-486-21076-6
Notwithstanding the title, it is a great beginning, intermediate, and advanced guide to RF theory and troubleshooting.

The Circuit Designer's Companion
Williams, Tim
Newnes, 1991
ISBN 0-7506-1756-X
For those interested in the proper design of analog circuits without the need for immense mathematical knowledge.

Communication Electronics
Frenzel, Louis
McGraw-Hill, 1995
ISBN 0-02-801842-7
One of the finest books in practical and current RF circuits and theory.

Computer Technician's Handbook
Margolis, Art
McGraw-Hill, 1990
ISBN 0-8306-3279-4
No RF, but plenty of useful digital troubleshooting information for any technician.

Electronic Circuits and Applications
Grob, Bernard
McGraw-Hill, 1982
ISBN 0-07-024931-8
Covers RF, digital, power supplies, audio circuits, etc.

Electronic Communications Systems
Dungan, Frank
Breton Publishers, 1987
ISBN 0-534-07698-X
Overview of communications in general, including RF, microwaves, telephones, satellites, etc.

The Illustrated Dictionary of Electronics
Turner, Rufus
Tab Books, 1982
ISBN 0-8306-1366-8
Fantastic reference.

Lenk's RF Handbook
Lenk, John
McGraw-Hill, 1993
ISBN 0-8306-4560-8
Covers the field of RF troubleshooting and operation.

Radio Operator's License Q&A Manual
Kaufman, Milton
Hayden Books, 1987
ISBN 0-8104-0666-7
*One of **the** books on everything RF. Targeted to those interested in obtaining the prestigious FCC GROL license.*

INDEX

Symbols

3T Circuit 229
3Ts 229
723 Adjustable Voltage Regulator 229

A

A1 Communications 129
AB Biasing 33
Absorption Trap 247
Absorption Wave Meter 247
AC Alternation 17
AC Circuit 3, 4, 21
AC Current gain 245
AC Half-Wave Current 17
AC Voltage Divider 225
Active Bandpass Filter 25
Active Clipper 233
Active Device 247
Active FET Probes 189
Active Filter 19, 23
Active High-Pass Filter 25
Active Phase-Splitter 49
ADC 115
Adjustable Frequency Divider 68
AFC 63, 94, 219
AGC 107, 213, 219, 223
AGC Bias Line 221
AGC Circuits 170
AGC Connection 43
Air Core 37
Air-Core Inductors 26
Aircore Transformer 26
ALC 213
Algorithms 233
Alloy Bar 111
Alpha Cutoff 61
Alpha Cutoff Frequency (f_{ab}) 247
Alternating Current Circuits 3
Aluminum 211
AM 83, 87, 105, 113, 115, 117, 137, 181, 213
AM Carrier 75
AM Components 95
AM Disadvantages 82
AM Fundamentals 71

AM Modulation 141, 181
AM Modulation Percentage 181
AM Modulation Percentage Test 181
AM Power 245
AM Power Measurement 77
AM Receiver 7, 77, 89, 170
AM Receiver Troubleshooting 170
AM RF Modulator Stages 33
AM Superheterodyne Circuits 77
AM Superheterodyne Receiver 79
AM Transmitters 79, 105
AM Waveform 82
AM/FM 186, 187
AM/FM Output 186
AM/FM Output Power Test 186
AM/FM Transmitter 187
AM/FM Transmitter Frequency Test 187
AM/PM 137
AM/PM States 121
Amp/Detector/Limiter 101
Amplification 9
Amplified Carrier Oscillator 131
Amplified Output Voltage 13
Amplifier 3, 27, 33, 34, 41, 79, 131, 139, 143, 166, 183
Amplifier Bandpass 183
Amplifier Bandpass Test 183
Amplifier Biasing 29, 33
Amplifier Circuit Configurations 27
Amplifier Circuits 27
Amplifier Coupling 35
Amplifier Efficiency 245, 247
Amplifier Frequency Limitations 59, 139
Amplifier Operation 31
Amplifier Self-Oscillation 166
Amplifier Self-Oscillation Troubleshooting 166
Amplifier Supply Voltages 236
Amplitude 63, 83, 85, 105, 117, 163, 169
Amplitude carrier 119
Amplitude Distortion 41, 169
Amplitude Modulation (AM) 71, 75, 95, 103, 117, 131
Amplitude Variations 71, 83, 97
Analog Information Transfers 115
Analog Output 119
Analog Pulse Modulation (APM) 115, 117

Analog-To-Digital Converter (ADC) 115
Angle (FM) Modulation Methods 3
Angle Modulation 83
Angle-Of-Radiation 247
Anode Current 17
Anode-To-Cathode Voltage 17, 19
Antenna 89, 109, 121, 135, 147, 149, 253
Antenna Current 75
Antenna Matching Networks 155
Apparent Power 244, 247
Application Specific Integrated Circuit (ASIC) 247
Atoms 9
Attenuation 141, 181
Attenuators 190, 247
Audio 109, 159
Audio amplifier 7
Audio amplitude 95
Audio frequency filtering 23
Audio gain 79
Audio input 95
Audio Modulation Frequency 87
Audio Modulator 82
Audio Power Amplification Stage 89
Audio Signal 89, 95
Audio Stages 77
Audio Wave 75
Audio-Derived AGC 107
Auto 191
Auto Transformer 247
Autodyne Converters 217
Automatic Frequency Control (AFC) 63, 94, 219
Automatic Gain Control (AGC) 43, 219
Automatic Level Control (ALC) 213, 247
Automatic Noise Limiter (ANL) 247
Automatic Power Control (APC) 129
Automatic Volume Control (AVC) 219
Autopatch 247
AV 54
Avalanche 11
Axial Leads 248

B

B cutoff frequency 15
Back-to-Back SCRs 17
Backplane 248
Bad Part 161
Bad Stage 160

Balance 151
Balanced Lines 151
Balanced Modulator (BM) 103, 107, 109
Balun 26, 157, 248
Band utilization 125
Band-Rejection 21
Bandpass 19, 21, 25, 248
Bandpass Filter (BPF) 21, 25, 75, 113, 119, 123
Bandspreading 248
Bandstop 19, 21, 23, 25
Bandwidth 23, 25, 75, 85, 119, 137, 248
Barrier Potential 9
Barrier Voltage 9
Barritt Diodes 141
Base 13, 27
Base Circuit 13
Base Current 13
Base Resistor 33, 34
Base-Biased Emitter Feedback, 33, 43
Base-Emitter Junction 13
Base-to-Emitter Capacitance 59
Base-to-Emitter Voltage Drop 33
Baseband 83
Baseband Signal 71, 81, 105, 248
Basic Safety Rules 159
Basic Troubleshooting 160
Basic Troubleshooting Steps 160
Batteries 13, 225
Battery Ampere-Hours 248
Battery Saver 101
Beam Power Tubes 57
Beam Width 248
Beat-Frequency Oscillator (BFO) 107, 131
Beats 215
BFO 105, 107, 131, 135
BFO Frequency 107
Bias 17
Bias Voltage 9
Biasing 33
Bidirectional Triggers 17
Binary 119
Binary Phase-Shift Keying (BPSK) 119
Binary-Coded Decimal (BCD) 248
Binocular Broadband Transformer 26
Bipolar (BJT) Transistor 13
Bipolar transistor 15, 95
Bit 248

Bleeder Resistor 225
Blocked-Grid Keying 131
Blocking Interference 248
Bootstrap Amplifiers 45
Bootstrap Circuit 45
BPF (Bandpass Filter) 75, 101, 119, 123
BPSK 119, 121
Bracketing 160
Breakover 19
Breakover Voltage 17
Bridge Rectification 225
Bridge Rectifier Diode Packages 11
Bridge-Connected Diodes 109
Broadband Power Amplifiers 248
Broadband Transformer 26
Broadband Transmission-Line Transformers 26
Buffer 95, 101
Buffer Amplifier 249
Buffer Register 249
Buncher 143
Bursts 99
Butterworth Characteristic 249
Bypass Capacitor 249
Byte 249

C

Capacitance 11, 63, 211, 249
Capacitance and Inductor Multipliers 249
Capacitance Variations 13
Capacitive 4, 97
Capacitive circuit 3
Capacitive coupling 37
Capacitive reactance 3, 23, 25
Capacitor 3, 19, 21, 23, 25, 95, 165, 179, 244
Capacitor Coupling 35
Capacitor Out-Of-Circuit Test 179
Capture Effect 87, 89
Carrier 71, 82, 85, 103, 107, 109, 111, 125, 249
Carrier Frequencies 97, 127
Carrier Null 85
Carrier Oscillator 97, 105
Carrier Shift 249
Carrier Suppression 111
Carrier Wave 71
Carrier-Operated Relay (COR) 135
Cascode 19

Cascode amplifiers 45
Catcher 143
Cathode 27, 57, 143
Cathode Depletion 175
Cathode Ray Tube (CRT) 71
Cathode-Bias 57
Cat's Whisker 139
Cavity Resonators 135, 141, 145, 146
CC Amplifiers 245
Cell Phone 129
Cell Sites 127
Cellular Communications 127
Cellular Radio 113, 127
Cellular Transceiver 127
Center Frequency 65, 83, 89, 191
Center-Tapped 47
Center-Tapped Input Transformer 49
Center-Tapped Output Transformer 49
Ceramic Crystal Transducer 111
Ceramic Filters 25, 249
Ceramic Ladder Filter 25
Ceramic-Chip Capacitors 211
Channel 17, 99, 119, 121, 249
Channel Guard 97
Channel Interference 87
Channel Resistance 15
Characteristic Impedance 249
Charge Carriers 15
Charges 13
Chebyshef Characteristic 250
Chip Quick□□TM 207
Chip Select 208
Chirp 133
Choke Action 250
Chokes 26
Circuit 3, 4, 23, 213, 243
Circuit Generated Noise 42
Circuit Protection 231
Circuit Sharing 160, 250
Circuits 9, 11, 165, 233
Circularly Polarized Antennas 123, 149
Circulating Currents 23
Clampers 233, 235
Clarifier Control 107
Class A 31
Class A Biasing 33
Class A Single-Ended Amplifiers 31
Class AB Operation 31
Class B Biasing 35
Class B Operation 31

Class B Push-Pull 47
Class C 33, 63
Class C bias 33, 34
Class C Amplifier 97
Class C Collector Modulation 71
Classes of Operation 31
Click 187
Clippers 233
CMOS Logic 250
CMRR 54
Coax 151, 153
Code 131
Code Division Multiple Access
(CDMA) 127
Code Generator 125
Coil 21
Collector 7, 13, 27
Collector Bias 43
Collector Current 13
Collector Inductor 37
Collector Load Resistors 53
Collector Modulation 71, 81, 82
Collector Modulator Circuit 81
Collector-Emitter 33
Collector-Feedback Bias 33, 34
Collector-to-Base Breakdown Voltage
13
Collector-to-Base Capacitance 63
Collector-to-Emitter Capacitance 59
Collinear Array 151
Colpitts 65
Colpitts Crystal Oscillator 67
Colpitts Oscillators 63
Colpitts-Based Oscillator 94
Combination 59
Common Emitter Amplifier Gain
Calculations 245
Common Mode Rejection Ratio 246
Common Tests 175
Common-Base 27, 60, 61
Common-Collector 27, 60
Common-Collector Amplifier 29
Common-Emitter 27, 60, 61
Common-Emitter Amplifier 29
Common-Emitter Circuit 33
Common-Emitter Transistor Circuit 13
Common-Mode Rejection (CMMR) 53
Common-Mode Signal 51, 54
Communication Applications 113
Communication Satellites 121
Communication Techniques 113
Communication-Service Monitor 250

Compandoring 250
Comparator Amplifiers 233
Compensation 59
Complementary Symmetry 49
Complex Wave 250
Component Values 23
Components 26, 139, 159, 169, 173
Compression 250
Conductive Loop 143
Conductor 3
Configurations 235
Connections 169
Constant Amplitude Pulses 115
Constant-Current Amplifier 111
Constant-Current Source 235
Constant-Current Source Circuits 235
Constant-K 19
Constant-K Low-Pass Filters 19
Constellation Diagram 121
Continuous Wave (CW) 129
Continuous-Tone Coded Squelch
System (CTCSS) 97
Control Gate 43
Control Grid 54, 55
Controller 135
Conversion 79
Conversion Receiver 79
Converter 250
Converter (DC-to-DC Converter) 250
Copper 139
Copper Losses 250
COR 135
Cosmic Noise 42
Counter-EMF 3
Counterpoise 149
Coupling 37, 191
Coupling Capacitor 251
Cross Needle Meters 199
Cross-Modulation 251
Crossover Distortion 169
Crosstalk 251
Crowbar Circuit 231
Crystal 25, 26, 65, 111
Crystal Controlled 94
Crystal Filters 25, 111
Crystal Frequency 65
Crystal Ladder 111
Crystal Ladder Filter 25
Crystal Lattice 9, 25, 111
Crystal Oscillator 63, 65, 101, 109
Crystal Oscillator Frequency 68
Crystal Oven 251

CTCSS 99, 135
CTCSS Encoder 99
CTCSS Equipped Transceiver 99
CTCSS Receiver 99
CTCSS Tone Generator 99
Current
3, 9, 15, 17, 19, 21, 55, 244
Current Controlled 13
Current Controlled Resistance 13, 141
Current Divider Formula 243
Current Gain-Bandwidth Product 15
Current Limiting Circuit 231
Current Output 15
Current Protection Circuits 17
Current Stream 9
Current Variations 3
Cutoff 54
Cutoff Frequency 251
CW 79, 105, 113, 117, 133
CW Carrier 87
CW Communications 129
CW Receiver 131
CW Signal 131
CW Spurious Responses 133
CW Transceiver 133
CW Transmitter 131

D

D-to-A Conversion 115
D/A converter 69
Damping Factor 246, 251
Darlington Pair 251
Darlington Pair Amplifier 51
Data Communication 113, 115
dB Meter 183
dB Sideband Level 181
DC 23, 107, 236
DC Amplifier 37, 51, 91
DC Audio Amplifiers 212
DC Bias 13, 55, 219
DC Bias Voltage 219
DC Control Voltage 99
DC Coupling 37
DC Current 3, 17
DC Current Gain 245
DC Restorer 235
DC Transistor Bias Voltage 13
DC-to-DC Converter 250
DDS 69, 109
DDS Frequency Synthesis System 69
De-Emphasis 87, 99
Decaying Sinusoidal Wave 61

Decibel Power Calculations 5
Decibels (dB) 4, 7
Decoder 119
Decoding 251
Degenerative Feedback 23, 251
Delayed AGC 109, 221
Delta-Match 155
Demodulation 71, 77, 83
Demodulator 91
Demodulator Probes 187
Demultiplexer 115
Denominator 7
Depletion Mode 15
Depletion Region 9, 11, 15
Depletion-Mode MOSFETs 15
Desensing 251
Desensitization 135, 146, 251
Desoldering 206
Despread signal 127
Detection 9
Detector 89, 91, 139
Deviation Ratio 89
DIAC 17, 251
Die Form 251
Dielectric 11, 145
Difference Frequency 135
Differential Amplifiers 51, 109
Differential Input 53
Differential Output 51
Differential Pair Amplifier 51
Differential Phase-Shift Keying
    (DPSK) 119
Digital Circuits 9, 71
Digital Data 115, 125
Digital Electronics 207
Digital Input 119
Digital Pulse Modulation (DPM) 115,
    137
Digital Radio 119
Digital Signal 119
Digital Signal Processing (DSP) 231,
    252
Digital Troubleshooting 207
Digital-to-Analog Conversion (DAC)
    115, 119
Diode 9, 94, 95, 145, 179, 223
Diode Clipper 233
Diode Detector 77
Diode Out-Of-Circuit Test 179
Diode Switch 236
Diode Terminals 9
Diode-Ring 109, 217

Diode-Ring Mixer 217
Dipole 147, 149
Dipole Antenna 149
Direct 93
Direct Conversion 251
Direct Coupling 37
Direct Current Amplifiers 51
Direct Digital Synthesis (DDS) 61, 68,
    69
Direct FM 83
Direct Modulation 94
Direct Waves 157
Direct-Sequence 127
Direct-Sequence Spread Spectrum
    (DSSS) 125, 127
Directional Coupler 129, 252
Directional Single-Needle 199
Directors 151
Disc Operating System (DOS) 252
Discriminator 94
Dish Antenna 151
Distortion 41, 97, 105, 169
DMMs 171
DMOS 252
Doping 9
Double-Conversion Circuits 123
Double-Ended 45, 47
Double-Ended Amplifiers 47
Double-Sideband Suppressed Carrier
    (DSB) 103
Double-Tuned 89
Double-Tuned Transformer 252
Double-Tuned Transformer Coupling
    39
Doublers 97
Doubly-Balanced Mixer 215, 217
Down-Conversion 215
Downlink 121, 252
Downlink Antenna 123
DPSK 119
Drain 7, 15, 17
Drain Current 15
Drain-to-Source Voltage (VDS) 15
Drift 65
Driven Array 151
Driven Element 149
Driver 82
Driver Stage 252
DSB Input Signal 111
DSSS Receiver 127
DSSS Transmitter 127
DTMF 129, 135

Dual-Conversion 105
Dual-Gate MOSFET 17, 217. 221
Dual-Gate MOSFET amplifiers 43
Dual-Tone Multiple Frequency
    (DTMF) 129, 135
Dummy Load 159, 252
Duplexer 129, 135, 146, 252
Duration 117
Duty Cycle 246, 252
Dwell Time 125
Dynamic Instability 173, 252
Dynamic Range 252
Dynamic Resistance 11

E

Ear 4
Earth Station Antennas 121
Earth Station Receiver 121
Earthbound Transmitter 121
Eddy Current Loss 26
Eddy Currents 252
Effective Radiated Power (ERP) 173,
    253
Effective Value 243
Electrical Impulses 25
Electrical Spark 17
Electrodes 25
Electrolytic Capacitors 211
Electromagnetic Interference (EMI)
    209, 252
Electromagnetic (RF) Wave 149
Electromagnetic Relay 133
Electromagnetic Wave 149
Electron Flow 9
Electron-Hole Pairs 9
Electronic Switches 236
Electronic Television News Gathering
    (ENG) 139
Electronic Troubleshooting 160
Electronics Industries Association
    (EIA) 252
Electrons 7, 9, 13, 17, 26, 145
Electrostatic Discharge (ESD) 205
Emitter 13, 27
Emitter Current 13
Emitter Follower 53
Emitter Resistor 33
Emitter-Base Barrier Voltages 13
Emitter-Feedback Biasing 34
Emitter-to-Collector Current 13
Encoding 252
End-Fire Array 151

Energy 21, 109
Enhancement Mode 15
Enhancement-Mode MOSFET 17
Equivalent Series Resistance (ESR) 231
ERP 245
Eutectic 207
Exciter 253
External 34
External Bias 13
External Bias Voltage 9
External Noise 42
External Signal 13
External Signal Voltages 13
Externally Generated Noise 42

F

F-Connector 253
Fading 82, 103, 253
Fan-Out 253
Faraday Screens 175
Faraday Shield 253
FCC Type Acceptance 253
FDM 113, 121
FDM Transmitter 113
FDM Wideband Receiver 115
Feedback 145, 166, 227
Feedback Amplifier 253
Feedback Network 25
Feedback Oscillations 166
Feedthrough Bypass Capacitor 253
Ferrite 26, 37
Ferrite Material 26
FET 7, 34, 94, 95, 97, 145
FET Amplifier Gain 245
FET-Based Phase Modulator 95
Fiberglass 145
Fiberoptic Cable 139
Field Strength, Antenna 253
Filter 19, 21, 23, 25, 109, 129, 225
Filter Designs 213
Filtering 225
Final 82
Final Amplifier 26
Final Power Amplifier 123
Fixed Frequency Crystal Oscillator 105
Fixed-Bias 59
Flat Line 155
Flattopping 41
Flexure-Mode Mechanical Filter 111
Flip-Flops 253

Flux Leakage 26
Flux Lines 26
Flywheel Effect 31, 33, 47, 253
FM 83, 87, 113, 115, 117, 135, 137, 183, 213
FM Broadcast 87
FM BW 245
FM Detectors 91
FM Deviation 183
FM Deviation Ratio 245
FM Deviation Test 183
FM Deviation-Meter 183
FM Fundamentals 83
FM IF Frequency 91
FM Modulation 83
FM Modulation Index 245
FM Receiver 7, 89, 171
FM Receiver Troubleshooting 171
FM Transceiver 99
FM Transmitter 93, 97, 99
FM Transmitter Power 85
FM Wave 89
FMing 173
Forward AGC 221
Forward Bias 9, 13
Forward Breakover Voltage 17
Forward Currents 11
Foster-Seeley Discriminator 89
Four Layer Diode 17, 19, 253
Frequency 21, 23, 25, 26, 68, 83, 85, 91, 94, 105, 131, 169
Frequency Applications 23
Frequency Compensation 59
Frequency Conversion 215
Frequency Converter Stage 160
Frequency Counter 201
Frequency Determining Network 23
Frequency Deviation 83, 87, 89
Frequency Distortion 41, 169
Frequency Divider 68
Frequency doublers 97
Frequency Drift 173
Frequency Hopping (FH) 125
Frequency Hopping Spread Spectrum 125
Frequency Limitations 59
Frequency Modulated Signal 85
Frequency Modulation (FM) 83, 85, 93, 95
Frequency Multiplication 94, 101
Frequency Multiplier 33, 97, 141

Frequency Multiplier Amplifiers 97
Frequency Response 26, 60
Frequency Reuse 123
Frequency Shift 254
Frequency Shift 173
Frequency Span 191
Frequency Stability 25
Frequency Swing 89
Frequency Synthesis (FS) 61, 68, 254
Frequency Synthesizer 61, 109, 125
Frequency Tolerance 254
Frequency Translation 121, 254
Frequency Variations 71
Frequency-Determining Crystals 101
Frequency-Division Multiplexing (FDM) 113
Frequency-Shift Keyer (FSK) 117, 125
FSK 117, 125
FSK modulator 125
FSK Signal 115
Full Duplex 254
Full-Wave SCRs 17
Full-Wave AM diode detector 77
Full-Wave Power Supplies 11
Full-wave Rectification 225
Function Generator 203
Functional Block Diagram 160
Fundamental 97
Fundamental Frequency 254
Fundamental Overload 254

G

Gain 23, 25, 107
Gallium Arsenide FET (GASFET) 141
Gallium Arsenide Varactor Diodes 141
GASFET 141, 254
Gate 15, 17
Gate Control Voltage 17
Gate Current 17
Gate Insulation 17
Gate Pulse 17
Glass Diode Packages 11
Gold Electrodes 26
Government Regulation 83
Grid 27
Grid Dip Meter 254
Grid-Leak 55
Grid-Leak Bias 57
Grid-Leak Components 57
Ground 21, 23, 169
Ground Clip 199
Ground Fault 254

Ground Loops 169
Ground Reflection 149
Ground Waves 157
Grounded-Cathode 27
Grounded-Grid 27
Grounded-Plate 27
Groundplane 149
Guard Bands 87
Gull-Winged SMD 207
Gunn Diodes 141, 143
Gunn Oscillator 141
Guy Wires 149

H

Half Duplex 254
Half-Wave Matching Stubs 155
Half-Wave Rectification 225
Hand-Capacitance 63
Hand-Off 127
Handy-Talkie 254
Hardware 254
Harmonic Crystal Oscillator 65
Harmonic Distortion 42
Harmonic Distortion Analyzer 255
Harmonic Generation 197
Harmonic Generator Frequency
    Interference 255
Harmonic Generators 33, 139
Harmonic Output 26
Harmonic Suppression 255
Harmonics 26, 41, 101, 175, 183, 197,
    254
Hartley 63, 94
Hartley Crystal Oscillator 67
Hartley Oscillator 63, 65
Hartley-Based Oscillator 94
HCDs 139
Heat Sinks 163
Helical 149
Helical Antenna 151
Helical Resonators 255
Helix 255
Heterodyne Frequency Oscillator
    (HFO) 133
Heterodyning 215
HEXFET 255
HF Transformer Coupling 37
HFO 133
High Frequency Transformer Cou-
    pling 37
High Image-Frequency 105
High Impedance 23

High Load Resistance 13
High Output Resistance 13
High Q tank 93
High Resistance 23
High Tank Impedance 23
High-Level Modulation 81, 82
High-Pass Filter 19, 23, 25
High-Q Crystals 19
High-Q Filters 145
High-Q Preselector 79
Higher Frequencies 23
Holder 25
Holder Capacitance 25
Holding Current 19
Holes 9
Hook-Up Wire 255
Horizontal Amplifier 195
Horizontal Antenna 149
Horizontally Polarized Antenna 149
Hot-Carrier (HCD) 139, 217
Hum 26
Hybrid Combiner 255
Hysteresis Loss 255

I

IC 9, 19, 91, 135, 165, 166
IC Detector 91
IC Differential Amplifier 109
IC IF Amplifiers 43
IC Op-Amps 53
IC Switching Regulators 231
IF 77, 79, 107, 109, 135, 217
IF Amplifier 43
IF Conversion 105
IF Strip 160
IGFETs (Insulated Gate FETs) 15
Image Frequency 79
IMDs 41
Impatt Diodes 141
Impedance 3, 4, 21, 23, 25, 26, 243
Impedance Coupling 37
Impedance Match 147, 243
Impurities 9
Inadequate Selectivity 169
Incidental AM (IAM) 93, 255
Incidental FM (IFM) 173, 193, 255
Indirect 93
Indirect FM 83, 95
Inductance 3
Inductance Values 26
Inductive 4, 97
Inductive Circuit 3

Inductive Coupling 26, 37, 43
Inductive Reactance 3, 244
Inductor 3, 19, 21, 23, 26, 225
Inductor Multipliers 249
Input 25, 167
Input Frequency 135
Input Level 190
Input resistance 13, 15, 245
Input Signal 111
Inputs 93
Insertion Loss 7, 26
Instantaneous Voltage 243
Insulator 15
Integrated Circuit SMDs 206
Integrated Services Digital Networks
    (ISDN) 137
Intelligence 71
Interference 169
Intermediate Frequency (IF) 77
Intermittent Problems 165
Intermittents 163
Intermodulation Distortion (IMD) 41
Intermodulation Distortion Products
    123
Internal Compensation 60
Internal Noise 15
Internal Resistance 11
Internally Generated Receiver Noise 7
International Telecommunications
    Union (ITU) 256
Interstage Matching Transformers 26
Interstage Transformer 26, 37
Intrinsic Semiconductor 9
Intrinsic Silicon 9
Inverter 255
Inverting Op-Amp 53
IREDS 255
Iron-Core Transformer 37
Isolator 255

J

JFET parameters 15
Junction 9
Junction Field Effect Transistors
    (JFETs) 15
Junction Tetrode Transistor 143, 256
Junction Transistor 13

K

Key Clicks 133
Keying Transistor 131

Klystrons 143
Knife-Edge Diffraction 256

L

L-Networks 39
L-Type 19
L-Type Constant-K Low-Pass Filter 19
Lagging 97
Laminated Iron-Core 37
LC Circuit 25, 243
LC Filters 25, 26
LC Matching Networks 43
LC Oscillator 61, 63, 65, 94, 219
LCD 69, 129
LCR Component Values 4
LCU 129
Lead Acid Cells 256
Lead Dress 163
Lead-Lag Network 68
Leads 173
Leakage Current 15
Leakage Value 9
Lecher Lines 256
LED 69, 129
Limiter 75, 93, 235
Linear 45
Linear IC 166
Linear IC Audio Power Amplifiers 47
Linear IC Troubleshooting 166
Linear Mixing 256
Linear Operating Region 15
Linear Single-Ended Class A 82
Link Coupling 236
Liquid Crystal Display (LCD) 256
Lissajous Patterns 197
Listen 160
Litz Wire 256
LO 105, 125, 129
Load 4
Load Resistor 15
Loading Capacitor 41
Loading Coils 256
Local Area Network (LAN) 125, 256
Local Oscillator Feedthrough 193
Locking 256
Log 193
Log-Periodic Array 151
Logarithm 5, 7
Logarithmic Response 4
Logarithmic Taper 256
Logic Gate 127

Logic Probe 205, 208
Long Range Communications 103
Look 160
Loose Coupling 39
LOS 157
Low Current Applications 11
Low Frequencies 23
Low Frequency AC 23
Low Frequency Transformer Coupling 37
Low Noise 17
Low Power 160
Low Reactance 23
Low Resistance 13
Low-Capacitance Probes 187
Low-Level Modulation 81
Low-Pass Filter (LPF) 19, 23, 26, 69, 93, 95, 101, 107, 168
Lower Frequency Voltages 23
Lower Sideband (LSB) 71, 109
Lowest Usable Frequency 157
LPF 101, 107
LSB 71, 109
LUF 157

M

M-Derived 21
Magnetic Field 3, 26, 109, 111, 145
Magnetic Lines 3
Magnetron 145
Main Line Current 23
Majority Carriers 9, 15, 17
Man-Made Noise 42
Marker 193
Master Oscillators (MO) 94
Matching Networks 39
Mathematics 4
Maximal Value 23
Maximum Allowable Collector Current 13
Maximum Current 17
Maximum Drain Current 15
Maximum Forward Current 11
Maximum Internal Junction Temperature 13
Maximum Usable Frequency 157
Maximum Zener Current 11
MCW 131
Mechanical Filters 111
Mechanical Impacts 139
Mechanical Resonant 111
Mechanical Vibration 25

MESFETS 141
Metal Can 25
Metal Film Resistors 211
mhos 15
Microcontroller 256
Microphones 257
Microphonics 257
Microprocessor 69, 257
Microstrip 145
Microwave 139, 155
Microwave Amplifiers 143
Microwave Bands 137
Microwave Bipolar Transistors 143
Microwave Communications 139
Microwave Detector 139
Microwave Diodes 139
Microwave Frequencies 139, 146
Microwave Integrated Circuits (MIC) 256
Microwave Oscillators 141
Microwave Region 151
Microwave Semiconductor 139
Microwave Signal 137
Microwave Transistors 141
Microwave Transmission Line 155
Microwave Tubes 143
Microwave Vacuum Tube Oscillator 145
Miller Capacitance 59
Miller Effect Capacitance 59
Minimal Value 23
Minority Carriers 9
Mixer 9, 33, 41, 119, 129, 139, 213, 215, 217
Mixer Circuits 11
MO 94, 95
Mobile Telephone Switching Office (MTSO) 127
Modulated Continuous Wave (MCW) 131
Modulating Audio 83
Modulation 71, 81, 83, 105, 183, 257
Modulation Circuits 11, 93
Modulation Frequency 213
Modulation Index 85, 87, 89
Modulator 119
Modulus 257
Monolithic Crystal 25
Monolithic-Crystal Filter 257
Monostable Multivibrator 91
Morse Code 79, 107, 131
Morse Key 131

MOSFET 17, 179, 215
MOSFET Out-Of-Circuit Test 179
Motorboating 169, 257
MTSO 127, 129
MUF 157
Multicavity Klystrons 143
Multilevel Encoding Schemes 119
Multimeter 189
Multiple Conversion Receivers 79
Multiple-Hop Transmissions 157
Multiple-Path Reception 257
Multiplexer 113, 115, 121, 137, 257
Multipliers 235
Multivibrators 257
Muting 223
Mutual Coupling 26
Mylar caps 211

N

N-Channel 15, 17, 177
N-Channel Device 15
N-Channel FET 177
N-Channel FET Out-Of-Circuit Test 177
N-Connector 257
N-Type Semiconductor 9
Natural Limitations 83
Natural Resonant Frequency 25
Negative 34
Negative Alternations 71
Negative Feedback 25
Negative Gate Voltage 15
Negative Potential 15
Negative Terminal 9
Negative Value 7
Negative Voltage 15
Network 258
Neutralization 55, 97, 167
Neutralized 43
No Power 160
Noise 41, 42, 83, 169
Noise Figure (NF) 258
Noise Immunity 87
Noise Voltages 89
Non-Inverting Configurations 53
Non-Inverting Input 25
Non-Resonant Filter 23
Non-Resonant Frequencies 23
Non-Resonant High-Pass 21
Non-Resonant Low-Pass 21
Non-Resonant Transformer 26
Non-Return to Zero (NRZ) 258

Nonconductive Tuning Tools 211
Nonlinear 45
Nonlinear Distortion 41
Nonlinear Mixer 103, 107
Normally-Off Device 15, 17
Normally-On Device 15
Notch Filter 21, 42, 258
NPN Doped Regions 13
NPN Transistor 34, 60, 221

O

Octave 258
Offset Oscillator 101
Offset-Null 54
Ohmmeter 179, 187
Ohmmeter Click Test 187
Ohm's law 4
Omnidirectional 149
ON device 15
On-Chip Converters 217
One Direction 83
Op-Amp 23
Operational Amplifier 23, 53
Optimum Coupling 39
Optimum Usable Frequency (OUF) 157
Opto-Isolator 258
Optocoupler TRIAC driver 19
Optoelectronic Devices 258
Original Equipment Manufacturer (OEM) 258
Oscillation 9, 63
Oscillator 41, 61, 63, 101, 107, 133, 139, 143, 145, 217
Oscilloscope 85, 105, 167, 171, 183, 195
Out-of-Phase Signal Input 97
Outer Shell 9
Output 21, 23, 25, 93, 167, 170
Output Current 13
Output Frequency 101, 135
Output Power 85
Output Signal 143
Output Tank 71
Output Tank Circuits 217
Output Voltage 15
Overcoupling 39
Overload Distortion 41
Overmodulation 75, 105, 183, 193, 213
Overmodulation Test 183

Overtone Crystal Oscillator 67
Overvoltage Protection 231

P

P Configuration 39
P Type 19
P-Network 155
P-Type Semiconductor 9
P-Type Semiconductor Current 9
P-Type Substrate 17
Packaging Single Crystals 25
Pad Attenuator 258
Padder Capacitor 258
Padder Trimmer 258
PAM 117
Parallel 19
Parallel (Tank) Resonant Filters 19
Parallel Amplifiers 51
Parallel Bandpass 23
Parallel Capacitance 244
Parallel Circuit 21
Parallel Damping Resistor 23, 45
Parallel Digital Word 115
Parallel Inductances 243
Parallel Inductive Reactance 244
Parallel Resistive/Reactive Circuit 244
Parallel Resonant Circuit 244
Parallel Resonant Frequency 21, 67
Parallel Resonant Mode 25
Parallel-to-Serial Converter 117, 125
Parameter 259
Parametric Amplifier 259
Paraphase Amplifier 49
Parasitic 149
Parasitic Oscillations 166, 167
Parasitic Suppressor 168
Parasitics 193
Part 161
Parts Failure 165
Parts Layout Diagram 160
Parts-Per-Million 259
Passband 19, 111
Passband Frequencies 25
Passive 23
Passive Divider Probes 189
Passive Filter 19
Passive LC Filter 19
Passive Low-Pass Filter 25
Passive RC Components 23
PCM 115, 117
PCM Digital RF Communication 117
PCM Modulator 117

PDM 117
Peak Envelope Power (PEP) 77, 105, 244, 259
Peak Inverse Voltage (PIV) 11, 259
Peak-Reading RF Wattmeter 186
Peaking Coils 59
Pentode 55
PEP 244, 259
Percent Modulation 75, 181
Periodic Waveform 259
Phase 91, 95
Phase Angle 3, 4, 185
Phase Angle Comparison Test 185
Phase Modulation (PM) 83, 93, 95, 259
Phase Relationship 82, 93
Phase Shift 4, 61, 91, 97
Phase Shifter 97
Phase Splitters 3
Phase Splitting 49
Phase-Comparator 68
Phase-Locked Loop (PLL) 61, 63, 68, 94, 219
Phase-Locked Loop Demodulator 91
Phase-Shift 67
Phase-Shift Keying (PSK) 115, 119
Phase-Shift Oscillator 67, 68
Phase-Splitter 47
Phasing 112
Phasing Harnesses 151
Phasing Method 111
Physical Capacitors 3
Pierce Crystal Oscillator 67
Pierce Varactor Oscillator 94, 101
Piezoelectric 25
Piezoelectric Crystal 65, 111
Piezoelectric Crystal Substrate 26
PIN Diodes 81, 141
PIN Modulator 141
Pinch-Off 15
Pinch-Off Voltage 15
Plastic Diode Packages 11
Plate 7, 27
Plate Modulation 71, 82
PLL 63, 68, 109, 125
PLL Chip Inputs 69
PLL Filter 69
PM 113, 137
PM Circuits 95
PN Junction 9
PN Junction Diode 9
PNP Doped Regions 13

PNP Transistor 60
PNPN Devices 17
Point-Contact Diode 139
Point-Contact Transistor 143
Point-to-Point Wiring 259
Polarity 9
Polarization 123, 149
Pole 259
Polystyrene 145
Port 259
Position 117
Positive 55, 143
Positive Alternations 71
Positive Control Pulse 17
Positive Gate Voltage 15
Positive Number 7
Positive Terminal 9
Positive Voltage 17
Potentiometer 259
Powdered-Iron 37
Power 5, 82
Power Amplifiers (PAs) 45, 119, 159
Power Attenuator 181
Power Audio Amplifier 82
Power Diodes 11
Power Dissipation 19
Power Dissipation Rating 11, 15
Power Factor 244
Power Line 26
Power Measurements 105
Power Ranges 9
Power Ratios 5
Power Supplies 225
Power Supply Circuits 213
Power Supply Regulation 245
Power Supply Unit (PSU) 259
PPM 117
Pre-Emphasis 87, 99
Pre-Mixing 69, 133
Premixer 259
Preselector 43, 79
Primary 26
Printed Circuit Board (PCB) 145
Private Line 97, 171
Procedures 161
Product Detector 103, 107, 131
Pseudo-Noise Generator (PN) 125
PSK 119
PTM 117
PTT 99
Pulse Modulation 115, 117
Pulse-Averaging Discriminator 91

Pulse-Coded Modulation (PCM) 115, 117
Pulse-Width Modulation 117
Pulsed Tone 99
Pulses 23
Push-Pull 45, 47, 51, 82
Push-Pull Amplifiers 35, 49
Push-Pull Audio Amplifiers 31
Push-Pull, Transformerless Input Power Amplifiers 49
Push-Push Doubler 97
Push-to-Talk Button (PTT) 99
PWM 117

Q

Q 23, 25, 146, 244
Q Point 260
Q Sections 155
Q-Point 219
QAM 121, 137
QAM Modulation 119
QPSK 119, 121
Quad Flat Packs (QFP) 207
Quadrature Amplitude Modulation (QAM) 117, 121
Quadrature Detector 91
Quadrature Phase 260
Quadrature Phase Shifter 111
Quadrature Phase-Shift Keying (QPSK) 119
Quarter-Wave Matching Stubs 151
Quartz Filters 25
Quasi-Complementary Symmetry 260

R

Radial Leads 260
Radials 149, 260
Radiation 149
Radiation Patterns 147
Radiation Resistance 260
Radio Communications 71
Radio Frequency Interference (RFI) 209, 261
Radio Frequency Signal 71
Radio Horizon 157
Radio Links 121
Radio Signal 77
Radio Spectrum 19, 93
Radiolink 137, 139
Radiolink Microwave Repeaters 137
Rails 260

Random Access Memory (RAM) 260
Random Values 7
Rate 83
Ratio Detectors 89
Ratios 191
RC Coupling 35
RC Filter 23, 25
RC Network 68, 95
RC Oscillators 63, 67
Reactance 3, 21, 23, 244
Reactance Modulators 94
Reactive Circuit 4
Reactive Components 3, 4
Reactive Element 95
Reactive Power 244, 260
Reactor 260
Read Only Memory (ROM) 261
Real Power 244, 260
Receive 135
Receive Signal Strength Indicator
  (RSSI) 129
Receiver 26, 41, 77, 119, 121,
  125, 135, 159, 168,
  169, 170, 185, 186. 219
Receiver Case 169
Receiver Frequency 186
Receiver Frequency Response Test
  186
Receiver Incremental Tuning (RIT)
  107
Receiver Interference 169
Receiver Sensitivity Test 185
Receiver Stage 170
Receiver Troubleshooting 168
Receiving Antennas 149
Receiving Equipment 103
Rectification 9, 11, 225
Rectifier 225
Rectifier Diode Specifications 11
Rectifier Diodes 11
Rectifier-Filter Variation 227
Reference Frequency 68
Reference Level 7, 190
Reflected Waves 153
Reflectometer 260
Reflectors 151
Reflex Klystron 143
Regeneration 169
Regenerative 143
Regenerative Feedback 260
Regenerative Repeater Transponders
  123

Register 260
Regulation 227
Regulator 225
Rejection 23
Relative Field Strength Meter (RFS)
  261
Relative Signal Strength Units 221
Relaxation Oscillators 19
Relay 131
Reluctance 260
Repeater 101, 113, 121, 135, 261
Residual FM 173
Resistance 21, 185
Resistance In-Circuit Test 185
Resistive Circuit 4
Resistive Components 3, 4
Resistive Element 95
Resistive Terminator 159
Resistive/Reactive Circuit 244
Resistor 4, 15, 23, 57, 159, 165
Resistor Array 261
Resistor-Capacitor (RC) 67
Resonance 21
Resonant 21
Resonant Antenna 147
Resonant Cavity 146
Resonant Circuit 21
Resonant Condition 21
Resonant Disks 111
Resonant Filters 19
Resonant Frequency 21, 23, 25, 243
Resonant Tank 95
Resonant Tank Circuit 21
Resonant Tank Impedance 21
Resonator 25, 261
Response 25
Response curve 23
Rest 65
Rest Frequency 89, 93
Resultant Phase Angle 95
Return-to-Zero (RZ) 261
Reverse AGC 221
Reverse Bias 9, 11, 94
Reverse Breakdown 11
Reverse Current Flow 9
Reverse DC Bias 141
Reverse Voltage 11
Reversed Biased 13
RF 23, 26, 77, 107, 159, 236
RF Amplifier 43, 129
RF Amplifier Stage 79
RF Carrier 95

RF Circuits 17
RF Connector 250
RF Design 26
RF Detection Circuits 11
RF Electronics 3
RF Matching Networks 39
RF Output Voltage 23
RF Parallel Power Amplifier 49
RF Power Density 245
RF Probes 187
RF Propagation 157
RF Receivers 77
RF Signal 77, 223
RF Spectrum 123
RF Transistor Amplifiers 26
RF Transistors 173
RF Transmission Line Noise 169
RF Transmission Lines 151
RFC (Radio Frequency Choke) 35, 236
Rheostat 261
Ringing 60, 261
RIT 107
RMS 243, 244
Rotor-Stator Air-Tuned Capacitor 211
RX/TX Selector 101

S

S/N (signal to noise) ratio 7, 43, 91
Safety 159
Sampled 231
Sampling Points 117
Sampling Scopes 261
Satellite 113, 121
Satellite Communications 121, 139
Satellite TV 139
Scan-Width 193
Scan-Width-Per-Division 191
Schematic 161
Schematic Diagram 160
Schmitt Trigger 115, 117, 261
Schottky Barrier Diodes 139
SCR 17, 179
SCR Out-Of-Circuit Test 179
Screen Grid 55
Second Harmonic 141
Selective Calling 97, 99, 261
Selective Fading 261
Selectivity 25, 43, 262
Selectivity Factor 261
Self-Bias 33, 34, 35
Self-Oscillation 166, 169
Self-Resonance 262

Semiconductor 9, 143, 165
Semiconductor Current 9
Semiconductor Diode 9, 13
Semiconductor Junction 7
Semiconductor Materials 11, 15, 60
Semiconductor RF Amplifiers 26
Semiconductor Switch 131
Sending 149
Sense 149
Sensitivity 7, 43, 262
Serial Binary Data 125
Serial-to-Parallel Converter 125
Series 59
Series and Parallel Resonant Circuits 25
Series Bandpass 21
Series Capacitance 244
Series Capacitive Reactance 244
Series Filters 19
Series Inductances 243
Series Inductive Reactance 244
Series Peaking Coil 60
Series Resistance 4
Series Resonance 21
Series Resonant Circuit 244
Series Resonant Frequency 25, 67
Series Resonant Mode 25
Series-Pass Regulators 227
Series-Pass Transistor Regulator 227
Series-Shunt 59
Service Literature 160
Servicing Block Diagram 160, 161
Shielding 171
Shockley Diode 19
Shorting Bar 155
Shot Noise 7, 42, 262
Shrink-Fit Tubing 262
Shunt 19, 59
Shunt Capacitance 59
Shunt Capacitors 25
Shunt Damping Resistor 23, 45
Sideband 75, 82, 85, 87, 103, 111
Sideband Attenuation 111
Sideband Cutting 169, 262
Sideband Filter 111
Sideband Production 87
Sideband Suppression 25
Siemans 15
Signal 13, 34, 105, 107, 127, 129
Signal Currents 139
Signal Diode 11, 262
Signal Distortion 15

Signal Generator 203
Signal Injection 160, 163
Signal Monitor 262
Signal Output 173
Signal Strength Indicator 221
Signal Switching 141
Signal Tracer 160, 161, 262
Signal-Bias 34
Signal-to-Noise Ratio (S/N) 121, 245, 262
Signal-to-Noise-to-Distortion 7
Significant Sidebands 85
Silicon 9
Silicon Anode Chip 139
Silicon NPN 143
Silicon NPN Transistor 13
Simplex Operation 101, 262
SINAD 7, 245
Sine Currents 3
Sine Voltages 3
Sine Wave 23, 93, 243, 244
Sine-Wave Current 3
Sine-Wave Oscillator Classifications 63
Sine-Wave Oscillators 61, 67
Single Chip 89
Single Conversion Receiver 107
Single Ended Amplifier 93
Single Sideband Suppressed Carrier (SSSC) 82
Single Tone Modulation 117
Single-Ended 45, 262
Single-Ended Amplifiers 31, 47
Single-Ended Bipolar Transistor Mixer 217
Single-Ended FET Mixer 215
Single-Ended Input Configuration 53
Single-Ended Output 53
Single-Ended RF Power Amplifier 47
Single-Ended SSB Linear Amplifiers 31
Single-Ended Transistor 217
Single-Filter SSB 111
Single-Sideband 103
Single-Sideband Suppressed Carrier 103
Single-Tone Modulation 85
Single-Tuned Transformer Coupling 39
Singly-Balanced 215
Sinusoidal Alternating Current 3

Sinusoidal Alternating Current Circuit 3
Skin Effect 262
Skip Zone 263
Skywaves 157
Skywave Propagation 89
Slew Rate 54
Slicers 233
Slider 155
Slope 89
SMA Connector 263
Small Outline Integrated Circuit Package (SOIC) 263
SMD IC 207
SMD Soldering 206
Smell 160
Snap Diodes 141
Sniffer Probe 167
Software 263
Solder Joints 169
Soldering 205
Solid-State CW Transceiver 133
Solid-State Switches 17
Sound Intensity 4
Source 15, 17
Source-Bias 34
Source-to-Drain Current 17
Space Noise 42
Space Waves 157
Space-Located Repeater Stations 139
Spaghetti 263
Speaker 91
Spectrum Analyzer 75, 167, 168, 171, 190, 197, 209
Speech 107
Speech Compression 213, 263
Speech Processing 213
Speech Processor 263
Splatter 75, 213, 263
Split 135
Spot Beaming 123
Spread Spectrum (SS) 113, 123, 125, 127
Spurious Emissions 263
Square 5
Square-Wave Analysis 185
Squelch 99, 223
Squelch Circuits 223
Squelch Gate 223
SS 125
SS signal 125

SSB 82, 103, 105, 113, 171, 186, 213
SSB AGC 107, 223
SSB Communications 25
SSB Detection 107
SSB Filter 111
SSB Frequency Test 186
SSB Fundamentals 103
SSB Generation 111
SSB Modulation 103, 105
SSB Output Power 105
SSB PEP 186
SSB PEP Transmitter Test 186
SSB Receiver 82, 103, 105, 107
SSB transceiver 105, 133
SSB Transmitters 111, 103, 109
SSSC 112
Stabistors 35, 49
Stacked-Series Output 49
Stage 160
Stagger Tuning 45
Standard AM 105
Standard Test Modulation 163
Static Electricity 17
Step Attenuator 181, 263
Step-Recovery Diode 141
Stop-Band 19
Strain Insulators 149
Stray Wiring Capacitance 59
Stripline 145
Stripline Filters 146
Stripline High-Q Filter 146
Superhet Receiver 105
Superheterodyne 131
Superheterodyne Principle 77
Superheterodyne Receiver 79, 215
Support Circuits 213
Support Components 89
Suppressor Grid 55
Surface Acoustic Wave (SAW) Filters 26
Surface Mount Device (SMD) 205, 263
Surface Mount Technology (SMT) 207, 263
Surface Waves 157
Surge Impedance 147
Sweep Generator 183, 263
Switches 139
Switching Power Supply Regulators 229
Switching Transistor 133

Switching Voltage 19
SWR 245
SWR/Wattmeter 171, 199
Symptoms 160

T

T-Network 39
T-Type 19
T/R Switch 135
Tank 21, 23, 95
Tank Circuit 21, 61, 93
Tantalum Electrolytics 212
Taps 37
TDM 115, 121
Teflon 145
Telegraph Key 133
Telemetry 263
Telephone Companies 139
Television Interference (TVI) 209, 265
Temperature Compensated Zener Diodes 11
Temperature Compensated Zeners 264
Temperature-Compensated Crystal Oscillators (TCXO) 265
Termination 264
Test Equipment 187
Tests 161, 175
Tetrode 55
Thermal Noise 264
Thermal Runaway 33, 264
Thermionic Emission 54, 145
Thermistor 264
Thermocouple 264
Thin Small Outline Package (TSOP) 265
Three Element Tube 54
Three terminal regulator (3T) 229
Three-State 208
Threshold Voltage 17
Through-Hole Soldering 205
Thyristors 17, 264
Time 243
Time Shift 4
Time-Delay Circuits 19
Time-Division Multiplexing (TDM) 113, 115
Time-Domain 75
Time/DIV 191
Timing 117
TNC Connector 264
Tolerances 11

Tones 99
Toroid Transformer 26
Toroidal Core 26
Toroidal Inductors 26
Toroidal Transformers 39
Toroids 26
Total Harmonic Distortion (THD) 42
Total Parallel Capacitive Reactance 244
Total Power Dissipation 13
Touch 160
TPTG 67
Tracking 264
Transconductance 15
Transducer 71, 83, 89, 115
Transformer 26, 37, 109, 211, 225
Transformer Coupling 37
Transformer Turns Ratio 243
Transient Intermodulation 264
Transient Voltage Suppressors (TVS) 264
Transistor 7, 13, 15, 82, 94, 95, 117, 145, 165, 168, 176, 177, 211
Transistor Circuit Configurations 60
Transistor Finals 173
Transistor In-Circuit Test 176
Transistor Out-Of-Circuit Test 177
Transistor Output Power 244
Transistor Ratings 13
Transistor-Transistor Logic (TTL) (T2L) Logic) 265
Transit Time 59, 139
Transit Time Problem 264
Translator 264
Transmatch 155
Transmission 103
Transmission Line 26, 145, 147, 151, 155
Transmit 133
Transmit Oscillator 125
Transmitter 41, 79, 95, 115, 121, 125, 135, 159, 168, 171, 175, 215
Transmitter Harmonics 175
Transmitter Problems 171
Transmitter Troubleshooting 171
Transmitter Tube 175
Transponder 121, 123, 264
Trap Antennas 265
Trapatt Diodes 141
Trapezoidal 183
Traveling-Wave Tubes (TWT) 123, 145

Tri-Statable 208
TRIAC 17
TRIAC triggering 19
Trimmer 265
Triode 54, 55
Tripler Amplifier 101
Tripler Varactor Frequency Multiplier 97
Triplers 97
Troposcatter 157
Troubleshooting
    159, 166, 168, 169, 170, 171, 175, 210
Troughs 75
Trunked RF Systems 265
Tube 7, 94, 173
Tube Amplifier Biasing 57
Tube Amplifiers 167
Tube Final AM Stage 82
Tube Types 55
Tunable Field-Strength Meter 183
Tunable Passive Filter 43
Tuned Circuits 37, 77, 145, 146
Tuned Filters 19, 21
Tuned Radio Frequency Preamp 43
Tuned-Plate Tuned-Grid Oscillator
    (TPTG) 67
Tuning Capacitor 41, 133
Tuning Indicators 221
Tunnel Diodes 141
Twin Cavity Klystrons 143
Twin-Antenna Repeater 135
Twin-Lead 151
Twisted Pair 265
Two-Tone Modulation 163
TX Clarifier 186
TX RIT 186

U

UHF 63, 103
UHF Connector 265
UHF Signals 101
Unbalanced 151, 215
Unbalanced line 153
Undesired Frequency 21
Unidirectional Conduction 11
Unijunction Transistor (UJT) 265
Unmodulated Carrier 71
Unpowered Amplifier Stage 167
Untuned Filters 19
Untuned Wideband Finals 26
Up-Conversion 215
Uplink 121, 265

Upper Sideband (USB) 71
Usable Frequency 157
USB 71, 109

V

Vacuum Tubes 9, 54, 143
Valence Electrons 9
Value 94
Varactor 11, 94, 101, 105, 141
Varactor Diode 11, 65, 95, 141
Varactor Modulator 94, 101
Varactor Tuning Diode 141
Variable Crystal Oscillator (VXO) 265
Variable Frequency Oscillator (VFO)
    63, 105
Variable Temperature Coefficient 211
Variable Video Filter 191
Varicap 94, 219
Varistor 265
Varying Input IF 91
VCO 65, 68
Vertical Antenna Radiates 149
Vertically Polarized Antenna 149
VFO 105, 109
VGS (gate-to-source voltage) 15
VHF 113
Vias 265
Virtual Ground 266
VMOS 266
Voice Amplitude Modulated Signal 75
Voice Signals 103
Voltage 3, 4, 5, 7, 21
Voltage Amplification 13
Voltage Amplifiers 42
Voltage Controlled Oscillator (VCO)
    94
Voltage Conversion 225
Voltage Dependent Resistor (VDR)
    266
Voltage Divider 23, 34
Voltage Divider, Emitter-Feedback
    Biasing 34
Voltage Divider Formula 243
Voltage Drop 9
Voltage Follower 53
Voltage gain 15, 43
Voltage Level 117
Voltage Multipliers 235
Voltage Reference 11
Voltage reference diodes 11
Voltage Regulation 11
Voltage Variable Capacitors (VVCs) 11

Voltage-Controlled Oscillator (VCO)
    63, 68
Voltage-Controlled Switches 17
Voltage-Divider Emitter Feedback 33
Voltage-Doublers 235
Voltage-to-Frequency Converter 68
Volume Adjustment 79
VZ Region 11

W

Wattmeter 186, 245
Wave Analyzer 266
Waveguides 151, 155
Wavelength 85
Wavetrap 101, 266
Wein-bridge 67
White Noise 7, 42
White Static Noise 41
Wideband 105
Wideband Amplifiers 45
Wideband Applications 26
Width 117
Windings 26
Wire Inductances 63
Wire Wound Resistors 211
Word 266

X

XOR 127

Y

Yagi 149
Yagi antenna 149
YIG 141
YIG crystal sphere 141
Yttrium-Iron-Garnet (YIG) 141

Z

Zener 109, 179
Zener Current 11
Zener Diode Specifications 11
Zener Diodes 11
Zener Shunt Path 227
Zener Shunt Regulator 227
Zener Shunt Voltage Regulator 227
Zener Voltage 179
Zener Voltage Out-Of-Circuit Test 179
Zero Beat 266
Zero-Crossing Detector 233
Zone of Silence 266

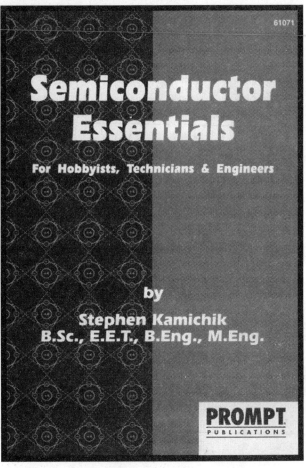